MW00622547

The Art of Military Innovation

The Art of Military Innovation

Lessons from the Israel Defense Forces

Edward N. Luttwak

Eitan Shamir

HARVARD UNIVERSITY PRESS

Cambridge, Massachusetts

London, England

2023

Printed in the United States of America

First printing

Library of Congress Cataloging-in-Publication Data

Names: Luttwak, Edward, author. | Shamir, Eitan, 1964– author.
Title: The art of military innovation : lessons from the Israel Defense Forces / Edward N. Luttwak, Eitan Shamir.
Description: Cambridge, Massachusetts : Harvard University Press, 2023. | Includes bibliographical references and index.
Identifiers: LCCN 2022062084 | ISBN 9780674660052 (hardcover)
Subjects: LCSH: Israel. Tseva haganah le-Yiśra'el. | Military art and science—Israel. | Military art and science—Israel—Automation. | Technological innovations—Israel.
Classification: LCC UA853.I8 L869 2023 | DDC 355/.03305694—dc23/eng/20230206
LC record available at https://lccn.loc.gov/2022062084

Contents

Introduction

THIS IS NOT A HISTORY OF THE ISRAELI ARMY, NOR A HISTORY OF Israel's wars. It is the record of an investigation that started with a simple question: why is it that the relatively small, relatively poor Israeli armed forces have long been exceptionally innovative?

Decade after decade, both under the urgent pressures of war, but also when threats were quiescent, the Israel Defense Forces (IDF), as officially translated in English, have kept innovating, originally largely by themselves, then in conjunction with the country's nascent military industries and fledgling research centers, both inevitably populated mostly by IDF veterans.[1] Over the years they originated new tactics in the air, at sea, for commando raids, and also for armored warfare—one particular 1973 tank battle was taken up as the model for a successful defense against numerically superior armored forces ("fight outnumbered and win") in subsequent US Army field manuals.[2]

At the higher, operational level of warfare, the Israelis have devised new schemes of combined action, often to take advantage of some new technology because they were so often the early adopters, but sometimes to make the most of entirely conventional forces.[3] In a major 1967 battle, Abu-Agheila/Umm Qatef (June 5–6, 1967), columns of foot infantry, tank battalions, massed field artillery, and descending airborne troops all attacked a well-entrenched, well-armed enemy by converging from different directions—seemingly a recipe for a friendly-fire disaster of epic proportions.

It was an attempt at a rock, paper, scissors–style synergistic victory: the airborne troops came by helicopter amid the Egyptian artillery batteries, whose crews were of course unprepared for hand-to-hand fighting; and the foot infantry did not attack Egyptian trench lines frontally but instead entered them by first slogging over supposedly impassable sand dunes to reach their unprotected starting points. Meanwhile the tanks, kept at bay by the antitank mines that protected the Egyptian trench lines, nevertheless forced the defenders to keep their heads down by frontal fire and threatened assaults, distracting them from the real threat of the Israeli foot infantry coming down their own trench lines. Instead of a friendly-fire debacle, it was the other side that disintegrated.

The striving to surprise the enemy by novel schemes of action, inevitably by accepting major and sometimes extravagant risks, characterized Israeli operations during the period of major conventional wars, which started in 1967 and ended in 1973 with a cease-fire that evolved into a "peace of the brave" with Egypt in the 1979 Peace Treaty. High-risk tactical innovations were pursued in ways large and also small in many commando operations on land, by sea, and by air. (In one, Super-Frelon helicopters delivered troops and equipment to a distant target beyond their maximum two-way range by returning empty to refuel before going out again to retrieve the soldiers.)[4] The alternative to such risk-taking has always been unacceptable to Israelis: straightforward, low-risk, frontal attacks would have cost tens of thousands of casualties during the years when the IDF fought the largest battles since the Second World War, with tanks and artillery pieces by the thousands, and jet fighters by the hundreds. Lives were saved by bold, risky innovations, not only in small commando raids but also on the largest scale. Perhaps the most extreme example of risk-taking was the assault-crossing of the Suez Canal in October 1973, which violated an elementary principle of war: Israeli forces did not even control their own side of the canal when they crossed over to the Egyptian side to attack the invaders from the rear. Its protagonist, the then divisional commander Ariel Sharon, was duly denounced as a reckless gambler by his colleagues—until he emerged victorious.

But the best-known Israeli innovations were neither tactical nor operational but rather technological, starting in the 1960s with the Gabriel, the West's first antiship missile cheaply developed by the then minuscule Israeli navy, when Israel was still an agricultural country of two million people, with little by way of industry, none of it advanced.[5] Arriving in time for the 1973 October war after a rapid development process, the Gabriel decided the outcome at sea by sinking nineteen Egyptian and Syrian vessels, with the loss of none.

That was the first of a sequence of true innovations, or *macroinnovations,* which were not merely new and improved versions of what already existed, but weapons or techniques that did not exist at all until then. In the 1970s macroinnovation was rather micro in form with the first of the small remotely piloted vehicles (RPVs), whose uses have kept expanding ever since, from overhead observation to different forms of attack and even transport; soon renamed unmanned aerial vehicles (UAVs), they then became today's ubiquitous drones. In what has become a global industry, Israel remains an important user and supplier of drones, exporting a variety of models from the smallest hand-launched types to sizeable aircraft that can execute long-range air strikes with significant weaponloads. Also in the 1970s, the Merkava tank was developed in successive versions while remaining the only main battle tank that diverges from the classic rear-engine configuration of every other main battle tank (and at sixty-five metric tons it is the heaviest, and seemingly the best protected).

In June 1982 Israel unveiled the first of a variety of air-launched armed decoys that magnified Israeli airpower, and while a new kind of tactical/strategic submarine is still a project wrapped in secrecy, the Iron Dome antirocket and antimissile system became globally famous in 2014 when it achieved unprecedented intercept rates against ballistic rockets, achieving even better results in the May 2021 fighting between Israel and Hamas in Gaza, when it reduced the impact of some 4,000 bombardment rockets to a handful of casualties. By then the Israeli-developed helmet-mounted display system had become the central innovation of the multiservice, multinational F-35 fighter family (it allows pilots to see right through the aircraft they are flying), while another innovation (not invention), the Trophy active protection system for armored vehicles, detects incoming antitank missiles and rockets with radar and engages them with self-forging munitions.[6] Now integrated with a competing Israeli system (Iron Fist), Trophy has been adopted by the US Army and others for their armored vehicles. In between those major innovations, there were many others, ranging from the reactive armor boxes that can be added to further protect armored vehicles, to the very concept of the multirole jet fighter, invented by the Israeli Air Force in the late 1950s when it was still insignificantly small by global standards, and its commanders mere youths.

This remarkable capacity for rapid innovation is facilitated by the peculiar organizational structure of the IDF, and while it is of course propelled by the security imperatives of Israel's situation, another factor is the educational impact of the IDF themselves, as the country's preeminent educational institution. Historically, national armies that conscript all the able bodied and therefore cannot be selective have long functioned as powerful educational institutions, teaching things as basic as personal hygiene and the difference between right and left feet as late as the Second World War in the US Army, along with an increasing variety of noncombat skills, everything from driving motor vehicles to the increasingly advanced use of computers.

That is true of the IDF as well, from the teaching of simple literacy and elementary arithmetic (for social reasons it accepts illiterate and innumerate conscripts) to the funding of advanced postgraduate studies for its officers. But the IDF adds a further element peculiar to itself: an intensely improvisational, can-do culture that can degenerate into overconfidence, but which is exceptionally open to innovation. What that means in practice is that no formal qualifications or position of authority are required to obtain a hearing for a new idea and even development funds if warranted—as quite a few Israelis at large, and some non-Israelis, have discovered over the years.[7]

[1]

Raising an Army under Fire

FROM THEIR BIRTH IN MAY 1948 THE ISRAEL DEFENSE FORCES (IDF) differed from other armed forces around the world. They were established from the first as a single service, with their fledgling ground, naval, and air units all under the same command, instead of the separate armies, navies, and air forces that existed elsewhere and persist to this day (except for Canada's much later and partly rescinded unification), sometimes with a fourth service, such as the US Marine Corps or Italy's carabinieri.

The IDF were and are also unique in conscripting women as well as men (though with exemptions more easily allowed). They have long relied on women trainers for every skill, including combat training, from the throwing of hand grenades to the firing of rifles, tank gunnery, and the operation of tube and missile artillery. Women therefore perform the combat instructor function prototypically performed by ultramasculine drill sergeant types in other armies. Many women soldiers perform administrative duties indoors, as many are combat trainers. Others have been volunteering to join combat units instead of just training others to fight, and some serve as air force pilots and naval combatants.

Originally, the conscription of women was simply a demographic necessity: a country that started off with a population of some 650,000 Jewish civilians when it was attacked by several Arab countries had to maximize its human resources by conscripting women as well. It was a simple matter of assigning as many noncombat tasks as possible to women, to release more men for combat. But over the years, it emerged that women provided not only numbers but also particular abilities, from the reduction of the fear factor in showing new recruits how to perform inherently dangerous actions such as throwing hand grenades, to quasi-maternal patience in imparting skills difficult to learn—both examples of recognizing rather than ignoring gender differences.

Yet another fundamental IDF innovation was inspired by Swiss military practice but carried much further: right from the start the IDF were established as a reserves-centered military force.[1] That was only theory in 1948 because the war started before any recruits had become trained soldiers. These would eventually become reservists, who could be recalled to active duty, but by the time of the 1956 war there were enough former conscripts to staff reserve formations that outnumbered the active-duty units, and that ratio increased further in 1967 and 1973.

That, of course, is the highly desirable advantage of reserve-centered forces: with any given population, they can be much larger than a conventional army staffed only by active-duty personnel. But a much less desirable consequence is that a reserve-centered army must be greatly dependent on advance warning to mobilize the bulk of its forces, a very tricky requirement that not even the best intelligence services can assure.

Of that, the best example was the failure of Israeli military intelligence to predict the October 6, 1973, surprise attacks by Egypt and Syria that were the start of full-scale offensives. In the bitter aftermath, that costly failure was blamed on the individuals in charge of Israeli military intelligence, who failed to revise their certitudes as the enemy forces facing them kept growing. But Israel's dependence on a reserves-centered army set up an impossible task for intelligence: at any given time, if a surprise attack is correctly predicted, the IDF mobilize, and the enemy calls off the attack. At that point, the indicators, both technical and human, including any agents-in-place within the enemy camp, that (correctly) predicted the enemy attack are discredited, while the misleading indicators that denied the danger of a surprise attack are validated. Repeated instances of crying wolf can blind the intelligence system—improving the enemy's ability to achieve surprise on the next occasion.

Exactly that sequence occurred in 1973. There had been indications of a coming war, but none occurred; then in May 1973 there was a large Egyptian buildup of forces on the Suez Canal facing the Israeli-held Sinai Peninsula, which triggered a large-scale mobilization of IDF reserve formations and individuals. That shut down much of the Israeli economy at great cost. When no Egyptian attacks materialized day after day, there was the demobilization dilemma: send the reservists home to resume their lives and return to work at the risk that they might have to be recalled to duty again in great confusion, or else wait a while longer as the costs pile up. In other words, a reserve-centered force offers great benefits, and inherent risks. Israel is a democracy—criticism in the media for the "superfluous mobilization" in May was remembered by decision-makers in October, as were American admonitions not to exacerbate the situation. They did not mobilize, and the enemy attacked.

Of all IDF innovations, the single-service structure is perhaps the most important. Its elementary economic advantages are obvious but perhaps not very significant for well-funded armed forces: yes, it is cheaper to have one headquarters structure, one set of uniforms, one set of administrative structures and practices and so on, but the real advantage of unified military forces is that unity favors innovation, as we shall see. Instead of the much-controverted and still-incomplete integration of armies, air forces, and navies that has slowly advanced in the United States as in most countries since the end of the Second World War, the IDF were born united right from the start, with no separate services, and of course no separate civilian ministries with different political ministers in charge. Those ministers, most notably the pre-1947 Secretary of the Navy and Secretary of the Army of the United States, or the British First Lord of the Admiralty, Secretary of State for War, and Secretary of State for Air, as with their equivalents elsewhere, had every reason to oppose the creation of an overall Defense Department or Ministry of Defence, which would make them subordinates of the one minister (or US secretary) of defense.[2]

It was the great predicament of the Jews of Palestine, whose state came under attack as soon as it was proclaimed on May 15, 1948, that they had no armed forces up and running, no army, navy, or air force ready for war to defend the new state. But in subsequent decades, it turned out that there was a hidden advantage in that most dangerous absence of an army, navy, and air force: it enabled the IDF to start anew as a single structure, achieving a unification still unachieved elsewhere.

Military institutions have no value unless they are sufficiently cohesive to generate and sustain the high levels of loyalty and dedication to duty necessary to fight in war, but those very same sentiments make military institutions highly resistant to change—and even if change is imposed on them, there may still be reversions as unchanged mentalities reassert themselves to restore what was there before the change. That is what happened with the Canadian Armed Forces—the only ones that tried to adopt the IDF's one-service structure. Until then as separate as their original British models, the Canadian Army, Navy, and Air Force were unified under the 1968 National Defence Act, which firmly stated that "the armed forces of Her Majesty raised by Canada consist of one Service called the Canadian Armed Forces or *Forces Armées Canadienne.*"[3]

It was hoped that unification would reduce the cost of socks as they could be bought in one color instead of three, with similar savings across a couple of million other items; and it was hoped that joint war planning and joint command in combat would be much easier if all concerned belonged to the same uniformed fraternity with the same vocabulary, habits, and procedures. The Canadians too had their bitter experiences of combat losses caused by interservice misunderstandings. At first it all worked out for the best with everybody in the same rifle-green uniforms of the unified Canadian Armed Forces. But atavistic identities persisted under the surface, and finally they won out: on August 16, 2011, forty-three years after their unification, the three "environmental" commands of the Canadian Armed Forces reverted to their original names: the Royal Canadian Air Force instead of Air Command, the Royal Canadian Navy instead of Maritime Command, and the Canadian Army (it was never "royal") instead of Land Force Command.

In further deference to tradition, uniforms reverted to their 1968 colors and patterns, complete with dress uniforms in red or blue, gold braid and all the rest.[4] Operationally this reversion is not supposed to mean anything at all, and indeed the separate services have not been formally reinstated—but what happened showed that strong institutional loyalties can override cold cost-benefit calculations, and with good reason: institutional memories and loyalties uphold morale and cohesion (i.e., esprit de corps), the two all-important if nonmeasurable intangibles that differentiate the relatively few armed forces that can actually fight armed enemies from the great majority that can only turn out on parade—and attack unarmed civilians. The issue of separate uniforms is merely symbolic of the truly problematic issues arising from separate weapon-development programs, separate training facilities, and separate, partially duplicative, administrations.

Institutional memories and institutional loyalties powerful enough to sustain combat morale will also impede innovation just as powerfully, if the particular novelty at hand collides with the missions, status, ethos, or self-images of those involved. That is where the single-service structure of the IDF really makes a difference in favoring innovation, simply because no one-service ethos or self-image stands in the way. Of this the clearest example is Israel's early leadership in unmanned aviation.

While pilots dominated the IDF Air Corps (Heyl Avir) as it then was, as much as pilots dominate almost all air forces, it was not a separate service.[5] Its commanders are subordinate to the single general staff of the IDF, and while pilots might have resisted the introduction of unmanned aircraft, the IDF General Staff did not. It was that organizational factor that allowed Israel to become the world leader in the design, development, and service introduction of unmanned aircraft, starting in 1970, when it was still a relatively poor, industrially undeveloped country of three million in all.[6]

Elsewhere, pilot-dominated air forces efficiently strangled unmanned-aircraft projects, even though the required technologies were so amply available that even toy manufacturers could and did offer remotely piloted aircraft with enough range and payload to be of some military use, right out of the box. Indeed, even now, over half a century later, with all manner of unmanned aircraft flying, including some that are very large and have intercontinental ranges, air forces everywhere continue to resist the adoption of unmanned aircraft (drones) in fighter and bomber roles, striving to reserve them for pilots, while confining unmanned aircraft to less heroic observation roles, except for some missile launches now and then. And that persists even though everyone understands that taking out the humans from the design of combat aircraft can drastically reduce their costs while adding much to their endurance and maneuverability (subtracting discretionary pilot control of course, but that is only important now and then). Because human g-force limits absolutely constrain the design of fighter aircraft, to avoid reversible grey-out, tunnel-effect, and blackout vision incidents, as well as G-LOC descents into unconsciousness and death, fighter aircraft valued for their agility and velocity are severely limited precisely in their agility and velocity by g-force limits that would not bother robotic aircraft at all, if their airframes are mechanically up to the task. In spite of this, not one unmanned fighter-bomber is in production, and even the future US intercontinental-range B-21 Raider heavy bomber meant for the delivery of both nuclear and nonnuclear weapons is to be manned as well as optionally unmanned, thereby adding robotic costs to the man-rating and the return-home costs, a very heavy price to pay to allow a few air force officers to pilot those aircraft, a choice that could only have been made by a pilot-dominated air force.

The impediments to innovation caused by otherwise praiseworthy and indeed essential service loyalties seem to affect navies even more than air forces—and that is readily understandable, given their much older origins. And the costs of naval loyalties have steeply increased in recent times, because they motivate the continued development and production of large and very large surface warships, in spite of their ever-increasing vulnerability to all manner of drastically cheaper weapons, including maneuverable reentry vehicles launched by ballistic missiles such as the Chinese DF-21D medium-range and DF-26 intermediate-range missiles, which could destroy any vessel they manage to hit with one descending warhead, including aircraft carriers.

All this was still in the unimaginable future when the State of Israel was inaugurated on May 14, 1948, the fifth day of the month of Iyar of the year 5708 in the Jewish calendar, facing a war already underway without an army, air force, or a navy to resist the armed forces of Egypt, Transjordan, Iraq, and Syria as well as armed bands both small and very large. It was then that the entirely new, single-service IDF was born, as a necessarily original invention because nothing like it existed anywhere in the world.

In the British forces—the ones best known to Palestinian Jews because so many had volunteered to join them during the Second World War—the Royal Navy, the Army, and the Royal Air Force each had its entirely separate administrative, cultural, and even political existence, and indeed differed not only in their appearance but, more profoundly, in their mentalities. Not surprisingly, interservice cooperation was difficult and sometimes simply impossible, and only improved slowly during the long years of war, from disastrous noncommunication at the start in the 1940 Norway Campaign to mere quarreling by 1944, even though Winston Churchill had established the world's first "Minister of Defence" position in 1940—wisely nominating himself for the post.[7] Even so, civil servants and budgets remained with the very separate secretary of state for war, first lord of the admiralty, and secretary of state for air right through the war—in other words, there was a defense minister, Churchill no less, but there was no actual ministry with its own staff or a single ministerial budget. A ministerial staff was provided in 1946, but the three service secretaries remained firmly in control until 1964, and it was only then that integration if not unification could begin—with no plans laid down to progress very far down that road.

The British model was therefore irrelevant for the one-service IDF, as was the brand-new American model established by the 1947 National Security Act, which instead of driving service unification, converted the US Army Air Forces into a separate service as the US Air Force, alongside the Army and the Navy, with its increasingly independent Marine Corps. The 1947 Act did also establish a single Department of Defense, but it did not abolish the separate service secretaries, with their secretariats and budgets. Hence successive secretaries of defense and their ever-growing staffs had to strive mightily over the decades to progress toward common planning and purchasing, with research and development even harder to unify.

As for the US structure for military command in war—the essential thing for Israel in 1948 with a war already underway in conditions of potentially catastrophic unpreparedness—it too was divided. The US Joint Chiefs of Staff, belatedly established in 1942 on the model of the British Chiefs of Staff Committee, could at most try to coordinate the separate planning and command of each service, because very literally, there was nobody actually in charge. Its head was neither a commander in chief (that being the president's prerogative) nor an executive chairman—that would only come forty years later with the sweeping 1986 reforms.[8] With no operational staff, and no military officialdom of his own, the first chairman of the Joint Chiefs of Staff, William D. Leahy, had more influence on broad strategy as the president's personal advisor than in the direct conduct of the war, because he had no effective authority over the chief of staff of the US Army, the US Navy's chief of naval operations, or the commanding general of the US Army Air Forces.

There appeared to be no valid model to follow for command structures—the Americans and British themselves kept saying at the time that they had won the war in spite of their command structures, not because of them, while too little was known of the much-admired Red Army.[9] Hence the Israelis of 1948 boldly disregarded all established practice and ignored all traditions to invent their own structure: one service, one general staff, one commander with the title "head of the general staff" (Rosh HaMateh HaKlali) under the immediate authority of the minister of defense, under the overall authority of the prime minister as head of the cabinet, or the cabinet as a whole in the gravest matters. While all other ex-British territories followed the prestigious model of the ever-glorious and newly victorious British armed forces,

Israel's leaders preferred to leap into the unknown with their own entirely original single-service IDF, a thing never before seen.[10] Thus, the Israelis were the first to venture on the path of service integration that others would follow in their own time, for their own reasons.

One important consequence, easily overlooked because it is a silent absence, is that the IDF never had to strive to harmonize different military services by maintaining joint command headquarters, a process that absorbs energies better used otherwise in perpetual strivings to maintain proper staffing balances between the services and a fair allocation of command slots. In contrast, the IDF have their institutional "jointness" built in, which makes it easier for different kinds of forces to cooperate logistically in peacetime, even if in war the soldiers on the ground, in combat aircraft above them, and in ships offshore will still have entirely different fields of vision, drastically different operational timelines, and weapons whose effective ranges vary from a few hundred line-of-sight meters to thousands of kilometers.

Hence coordinated air-ground fighting still needs much planning and much training, but at least in the IDF such efforts are not impeded by dysfunctional barriers between different institutions. It is to avoid that problem that the US Marine Corps fiercely holds on to its own "air wings" with their Marine-flown fighter squadrons, to provide close air support to Marines fighting on the ground, instead of relying on the US Air Force or indeed Navy pilots for that most difficult task. Even among fellow Marines there is still the need to coordinate different environmental perspectives and timelines, but communications are easier within the same military family and, critically, combat risks are more likely to be shared fairly between Marines above and below, with pilots taking risks to reduce risks on the ground and vice versa. That was famously demonstrated in the Korean War, when vastly outnumbered units of the First Marine Division fought their way south from the Chosin Reservoir from November 27 to December 13, 1950, with Marine pilots ignoring massed machine-gun fire to deliver their munitions with maximum accuracy to support fellow Marines on the ground.

Much the same thing occurred on October 6–10, 1973, when the Israeli Air Force sent its fighter-bombers to attack advancing Syrian forces on the Golan Heights even though they were protected by an abundance of Soviet-supplied antiaircraft missiles that could not be suppressed beforehand. Hugely outnumbered Israeli ground forces were thus able to resist, if only just, at the price of many pilots killed in their destroyed aircraft. As with the Marines, there was no institutional separation to diminish the urge of the pilots to help the soldiers on the ground, notwithstanding extreme risk.

Unified Structure and Innovation

When it comes to innovation, the benefits of the IDF's institutional unity are direct, simply because their research and development funds are not parceled out to the separate services that mostly use them to improve existing weapons vehicles and sensors—and especially their iconic platforms, as with the US Army's battle tanks, the US Navy's aircraft carriers and submarines, and the US Air Force's fighter jets and manned bombers. Such incremental innovation, soundly based on the remediation of specific shortcomings that have emerged over the years, or the straightforward replacement of old subsystems with new ones (as when an older jet engine is replaced with a new one that fits in the same volume), is much less risky than macroinnovation—that is, the research and development of something entirely new, which might fail entirely because of irremediable technology gaps, or simply because costs keep increasing with no end in sight.

Moreover, macroinnovation has another and weighty disadvantage: something really new will require the retooling of maintenance facilities and the retraining of their personnel, as well as the training of operating crews *ab initio*, which is costly and time consuming in itself. But macroinnovation offers one very great advantage over incremental innovation that can outweigh all its risks and costs: if the weapon or device is truly new, there will be no countermeasures nor counterweapons already in service with enemy forces to resist, counter, or outmaneuver the new capability. That absence suspends the entire predicament of war that makes it so hard to win battles and wars, namely the existence of opposing forces and opposing minds ready and waiting to observe and negate whatever is attempted.

Such a "countermeasure holiday" occurred on November 20, 1917, the first day of the Battle of Cambrai on the Western Front, when 378 Mark IV tanks of the British Army marked the first appearance of the macroinnovative battle tank in significant numbers. In the absence of antitank guns that had yet to be developed (low-slung guns, with high-velocity rounds to pierce armor), and in the absence of antitank mines—let alone the antitank rockets that came in the next war or the antitank missiles that arrived later still—those 378 Mark IVs flattened the forests of barbed wire that had defeated so many infantry assaults, drove right over the trenches that had harbored the riflemen and machine gunners that had previously cut down attacking infantry, and entirely negated both the bullets fired at them and the splinters of artillery shells with their steel-plate armor. Since the introduction to battle of the first tanks a year earlier during the Battle of the Somme, the British attacks had used only a few dozen tanks at any one time, and the Germans had been attempting to develop an antitank response. Of all the countermeasures employed, deploying light cannon to fire directly at the tanks had proven the most effective—and one such antitank battery delayed a British tank force at Cambrai considerably. But the few batteries available at the front could not defeat a massed attack by hundreds of tanks.

That is the reward of macroinnovation, which can win battles, even entire campaigns, if applied on a large-enough scale, by military leaders ready to take the risk of allocating important resources to the new and untried. Mostly they do not, because the really new weapon (as the tank was) will not enhance existing forces, will not affirm an existing way of war.

That is why the introduction of the first machine guns was resisted: none of the existing forces could use them. They were too heavy for the infantry to carry into battle, too clumsy to be mounted on a horse, and too piddling for the artillery that fired powerful explosive shells and not mere bullets. The same was true of the tank, a concept that the British Army refused to invest in, essentially because it threatened to displace the socially dominant horse cavalry and was bound to take away guns from the artillery while overshadowing the infantry. Those being the three branches that controlled the British Army, its leaders refused the idea—and so the first tank was developed by the Royal Navy at the insistence of Winston Churchill.

Simply because the IDF is not a military service or a federation of military services but rather a unified military body, it can accept macroinnovations and fund them, inevitably at the expense of the forces that already exist, because those separate forces with their separate identities are simply not in control. That is the ultimate explanation of the long list of Israeli military innovations. At the very start on May 15, 1948, some 650,000 Jews with a poor agricultural economy and hardly any industry could not develop or manufacture anything that exceeded the scope of a modest number of blacksmithing, welding, and mechanical workshops, with just a few veritable factories, none large, to produce textiles and clothing, canned food, and agricultural hand tools and such.

But at least the choice of what to develop and manufacture was easy indeed, because all the IDF had at the start were the meager caches of ill-assorted weapons secretly accumulated by the prestate militias, the dominant Haganah and its much smaller rival the Irgun, the result of a 1931 political split. The Haganah ("The Defense") enrolled men and women of all ages in its Heyl HaMishmar (guard corps), fit young people in its Heyl HaSadeh (field corps) and a few thousand in the select Plugot Maḥatz (Palmach—strike units).[11] The much smaller Irgun Tsvai Leumi (National Army Organization) only had a few organized units, while Lohamei Herut Israel (or Lehi; Fighters for the Freedom of Israel), aka the Stern Gang of the poet Yair Stern, did not exceed 300. Aside from odd lots of pistols, revolvers, some submachine guns, rifles of different calibers, shotguns, and a handful of machine guns, there was very little except for some trucks and buses protected with bolted-on steel plates.

The Jews could not legally import any weapons until British rule ended on May 15, 1948, and, because of a UN embargo, neither could their new state import weapons after May 15: even with the invasion of four Arab armies underway, it was nevertheless British policy, vigorously backed by the US government, to stop any weapons at all from reaching the belligerents, ostensibly to limit the violence, but actually to ensure the victory of the invading Arab armies that already had their equipment. That emerges very clearly from the authorized history of the British Secret Intelligence Service.[12] (The doings of intelligence services are rarely very consequential, but they do reflect actual policy aims more accurately than diplomatic declarations.)

Given what had very recently happened to millions of the Jews' coreligionists in Europe, the British stance toward the local Jews was more than harsh, but its motives were not deliberately malevolent. The British were merely being practical: at the time, they still had large military bases in the canal zone of Egypt they meant to keep, and imperial possessions east of Suez. They had trained and equipped the Egyptian Army of King Farouk that was then poised to drive to Tel Aviv; and the British had also funded, trained, and equipped the smaller Arab Legion of the Hashemite Kingdom of Transjordan, commanded by LTG Sir John Bagot Glubb ("Glubb Pasha") and his deputy Norman Oliver Lash, both British citizens, as were the thirty-five officers who commanded the Legion field units that crossed the Jordan to invade Palestine, attack Jewish settlements, and try to conquer Jerusalem.[13]

Iraq, much larger than Jordan and already oil rich, was also ruled by a Hashemite king installed by the British. Its army had also been equipped and trained by the British, and its government was dominated by Nuri al-Said, sturdy ally of British interests in Iraq until his murder a decade later.

With all those valuable British assets on one side, and some 650,000 oil-less Jews on the other, the British decision to support the Arabs and deny weapons to the new Israeli state was quite rational, as was the decision of the US secretary of state, five-star general, former army chief of staff, and future Nobel Prize winner George Catlett Marshall Jr. to back State Department officials who sided with the British against the White House in believing that President Truman's immediate recognition of Israel on May 15, 1948, was a great mistake that Israel's destruction by victorious Arabs would soon correct.[14] The British foreign secretary Ernest Bevin had already preemptively blamed impractical Zionist dreams for the inevitable massacre of the Jews.

It was therefore very unfortunate that Marshall's tenure (January 1947–January 1949) coincided almost exactly with the most critical phase of Israel's emergence. Though devoid of any personal animosity, let alone antisemitism, Marshall's opposition was absolute and relentless.[15] When Israel's envoy asked for an audience, he refused—he was much too busy with the nascent Cold War to waste any time over an ephemeral ministate that would soon be destroyed.

That was Marshall's prediction as an expert strategist, in which the newly established Central Intelligence Agency (CIA) fully concurred, and one he himself did much to bring about because US diplomats worldwide vigorously joined the British in preventing any weapons at all from reaching Israel.[16] Europe was then still littered with abandoned but still very usable weapons of all kinds, in all sorts of depots or under tarpaulins in the open, everything from rifles, artillery, and tanks to functioning or repairable combat aircraft. And Europe's impoverished postwar governments would eagerly have sold any weapons they had to the new Israeli state, which from May 15, 1948, had the legal right to buy anything it wanted. But as soon as word of a sale reached them, British and US diplomats would intervene, with their then-immense prestige, successfully in all cases but one: Czechoslovakia, a small country with world-famous small-arms factories eager for business.

That was indeed important because the newborn state could not hope to equip effective armed forces with the odds and ends that smugglers brought in—small batches of weapons of different calibers, often old and worn out, or missing parts. Nor could smugglers hope to bring in combat aircraft, armored vehicles, or field artillery—all too big to pass undetected. So extreme was the need for weapons once the Arab invasions started on May 15, 1948, that two antique French 65 mm howitzers (Canon de 65 M model 1906) with missing sights, which fired feeble ten-pound (4.4 kg) shells at a leisurely 330 m/s muzzle velocity, were viewed as weapons of strategic importance reserved for the highest-priority tasks, starting with the defense of the Degania sector on the Jordan River against the invading Syrian army on May 15–21, 1948.

Marshall's expert prediction might well have proven accurate had it not been for the coalition government of Czechoslovakia, which ignored Anglo-American pressures and sold weapons to Israel from its extensive stocks of weapons, both those produced by its sizeable and innovative military industries before 1938, and those produced under wartime Nazi direction thereafter.[17] Soon chartered transport aircraft flown by intrepid volunteer pilots delivered 34,500 German Mauser rifles that Israelis still call Czehi, along with 5,515 MG 34 medium machine guns, 500 ZGB 33 light machine guns, 900 ZB 53 medium machine guns, and more than a million rounds of ammunition.

The Czechs also had fighter aircraft to sell, sixty-one British-made Spitfire Mk IXs that had equipped the Free Czechoslovak Air Force squadrons of the Royal Air Force and were still first-line aircraft in 1948, and twenty-five locally made Messerschmitt Bf 109s (reuniting both protagonists of the Battle of Britain) and more dubious Avia S-199s.[18] In addition, the Czechs provided training for eighty-one pilots and sixty-nine ground crew, as well as an airfield for the transshipments to Israel. It was not much as compared to the inventories of the Arab armies, and it was impossible to airlift any of the locally plentiful armored vehicles in light transport aircraft, but the Czech shipments were enough to make all the difference by providing homogenous sets of small arms for IDF field units, leaving the odds and ends accumulated over the years for the local-defense units.

Not all the fighter aircraft survived the perilous transit (only possible at all because of a secret refueling airstrip made available by Tito's Yugoslav government), but enough did to allow courageous pilots to immediately go over to the offensive against the Egyptian Air Force. Equipped and trained by the British over the years, the Egyptians had already bombed Tel Aviv's central bus station on May 18, 1948, killing forty-one and wounding sixty, to this day a greater casualty toll than any subsequent Arab air attack in seven decades of intermittent wars.[19]

All-important as they were, the Czech shipments did not include any artillery or armored vehicles, both of which were essential to resist the invading Arab forces and then to go over to the offensive. That is how Israel's history of military research and development started, prompted by necessity rather than technological ambitions, with novel designs imposed by very severe technological limitations rather than any striving for originality for its own sake.

The Davidka mortar, the very first Israeli weapon developed from scratch, exemplified both characteristics. Three-inch mortars (actually 3.209 inches / 81.5 mm) were standard in the British Army, and the Davidka too had a base plate and a three-inch tube. But there was no available supply of three-inch bombs, and none could be produced in local workshops with the precision needed to avoid deadly in-tube explosions. The highly original solution was to make supercaliber bombs with a caliber-sized rod to propel them that could be safely projected from the mortar's barrel. With four times as much explosive as British three-inch mortar bombs, the Davidka made for very loud explosions, but lacking in both accuracy and range, it was more useful to frighten enemies than to attack their defenses. Only seven were made and they achieved little. As for armored vehicles, aside from two medium Cromwell tanks stolen by sympathetic British Army drivers during the final British withdrawal, and three defective US M4 Sherman tanks assembled from assorted wrecks left behind by the British Army, there were only improvised armored vehicles, made by bolting steel plates with firing slits onto trucks or buses, some with frontal rams to break through obstacles.

There was also an early anticipation of today's compound armor in the use of plywood, concrete, rubber, and even glass plate sandwiched between thin metal sheets. The different densities served to deflect bullets, while limiting total weight to avoid overtaxing engines and overloading chassis. But as soon as standard face-hardened steel plate could be imported, it was much preferred.[20]

Jeeps could not be armored but they could be armed, and the IDF equipped some with the formidable firepower of two MG 34 / 41 machine guns each firing 1,200 rounds per minute—ideal for fast raiding by providing short bursts of very intense fire to intimidate and suppress resistance. They equipped the Eighty-Ninth Commando Battalion formed and initially led by the subsequent general, chief of staff, and minister of defense Moshe Dayan, who was notably successful in gaining territory by hit-and-run raids.[21]

The jeeps arrived because even the otherwise very effective Anglo-American embargo could not prevent the import of entirely unarmed vehicles from the war-surplus dumps in Europe, even if they had originally been military vehicles. That category included thinly armored ten-ton US-made M-3 half-tracks that combined front wheels for steering with tracks in the rear for propulsion, and a protected front cabin with an open-topped rear cargo volume. They were to have a very long life in the IDF: while the US Army replaced all its half-tracks with fully tracked vehicles just as soon as it could, the IDF still used many in the 1982 Lebanon war, and some remained in use for another decade beyond that.

More than 3,000 half-tracks were imported by the IDF over the years, initially to serve as armored troop carriers but later adapted for many specialized roles: as command carriers with extra radios and a front winch; as weapon carriers for heavy machine guns, for 81 mm mortars, for locally made heavy 120 mm mortars, for twin 20 mm Hispano-Suiza HS.404 cannon, for rockets and mine-clearing Bengalore torpedoes, and finally for antitank missiles; and as combat-engineer vehicles, ambulances, and more. Half-tracks were modified for all these different purposes because they were cheap and abundant as compared to any other armored vehicle, and because the open-topped cargo area could readily be modified to accommodate weapon mounts or anything else, including armored cabins for the evacuation of the wounded. What happened with this particular vehicle through all its different modifications illustrates a fundamental aspect of IDF culture that persists vigorously even decades after the arrival of their first new weapons, delivered complete with replacement parts and specific maintenance tools: a proclivity for rehabilitating secondhand military equipment acquired in varied stages of disrepair, by repairing, retrofitting, and modifying what arrived till it becomes useful, either for the purposes originally intended, or for something else entirely.

During the years of large-scale wars from 1967 to 1973, the rewards of this proclivity were very substantial, indeed of strategic significance, because the IDF's arsenal was substantially reinforced with weapons captured on the battlefield. Thus US-made M48 Patton main battle tanks captured from the Jordanians were added to Israel's M48s originally delivered by West Germany when its army was reequipped with much more modern Leopard tanks. Over time, both were upgraded with new 105 mm guns and eventually reengined with powerful diesels in place of their original gasoline engines that caught fire all too easily. With those changes, Israel's M48s were almost as good as the newest M60s, which the IDF only acquired much later.

Many of the Soviet T-54/T-55 tanks captured in 1967 were cannibalized and repaired as necessary to equip new tank units, and then over time they were successively upgraded with new 105 mm guns, coaxial machine guns, radios, and sundry bits and pieces as the Tiran series of recycled tanks. With fewer or no modifications, excellent Soviet small arms, including the justly celebrated AK-47, were also transformed from battlefield loot into properly maintainable weapons that armed entire units, a process which was replicated with captured Soviet artillery, and particularly the 130 mm long-range gun.

The recycling of captured Soviet tanks acquired strategic importance because it allowed the IDF to keep up with the rapid post-1967 expansion of the Egyptian, Syrian, and Iraqi armored forces, at a time when the United States was only manufacturing thirty Patton M60s per month, and the United Kingdom offered its Chieftain tanks (initially codeveloped with Israel!) only to Iran and to Arab armies. In October 1973, Soviet antitank missiles and rocket-propelled grenades (RPGs) in huge numbers allowed brave Egyptian infantry troops to engage oncoming Israeli tanks, destroying some and immobilizing many more, to the point that Israel's armored forces were visibly shrinking just as Iraqi armored forces were arriving to join the fight. It was the Tirans as well as other captured Soviet tanks that allowed the IDF to achieve the organizational miracle of fielding a new armored division in ten days, with tank crews and all other essential personnel found by recalling older reservists, quickly retraining tank crews that had lost their tanks, and combing out armor and other combat soldiers who had drifted into administrative tasks.

The can-do, improvisational mentality originally instilled by the compelling necessity of rehabilitating abandoned equipment—everything from small arms to multiengine aircraft—by repairing, refitting, and retrofitting, has persisted through the decades even as the country acquired increasingly advanced laboratories and factories. It is manifest chiefly in a willingness to act quickly, accepting risks along the way—the exact opposite of the zero-risks mentality that slows weapon development to a glacial pace in Europe and the United States.

These days, with Israel sufficiently advanced to sell sophisticated weapons around the world, it must accept drastically different US methods when developing, producing, and modifying equipment for the United States. Shaped by countless regulations mandated by the US Congress in order to combat "waste, fraud, and mismanagement" in military procurement, the process is declaredly adversarial in imposing exhaustive documentation at each step, and "objective" testing: instead of quickly making proof-of-concept prototypes and testing them to uncover deficiencies that can quickly be fixed before the next test, engineers must take all the time needed—years sometimes—to perfect what they make even in the prototype stage, because it is tested by an outside test and evaluation entity that earns its keep by finding faults. They take their time to test exhaustively, covering all the just-in case possibilities in all weathers and all conditions, and then take more time to compile their reports.

It is only then that the development process can resume after any test failure—unless the project is cancelled on the basis of the report. Thus for example the United States remained without remotely piloted surveillance aircraft till Israeli ones were purchased in time for the 1991 Gulf War because the US-designed Lockheed MQM-105 Aquila, whose development started in 1972, was suspended in September 1985 (after thirteen years!) because the system failed 21 of 149 performance specifications, a number that itself proves frivolous excess in mandating outlandish requirements.

The phenomenal rapidity of the development of the Iron Dome antirocket system, from its first funding in 2007 through research and development, tooling and production, training, and deployment to its first successful combat use in April 2011, shows that the IDF's high-speed innovation culture is still in good working order: four years to develop a new missile is unheard of, but in this case there was also the extraordinary software that makes all the difference by only launching interceptor missiles against rockets projected to hit people or very valuable objects. That same example shows that the other innovation inheritance of 1948 is also still in place, because the phenomenally successful Iron Dome was not developed by the ground forces or the air force or the navy—none of which has its own research and development organization—but rather by the one and only research and development organization that serves the IDF as a whole.[22] That was just as well because the Iron Dome would never have been developed by ground-force officers who naturally prioritize the development of armored vehicles and other ground weapons, or naval officers with surface and submarine warfare on their minds, or by airmen whose solution for every problem is offensive airpower. Anything really new is unlikely to fit comfortably within existing service roles, so the necessary effort to research and develop the really new and bring it to operational status is unlikely to be forthcoming, as the chiefs of each service focus on their own priorities.

What would have happened to the Iron Dome in the absence of the one-service IDF is what actually *did* happen with the nondevelopment of US military space satellites, whose critical importance for the US Army, Navy, Marine Corps, and Air Force has never been disputed, but which none of the above was willing to develop with its own budgeted funds, preferring to use their funds to ameliorate their existing equipment. And this went on until the Soviet Union launched Sputnik, the first artificial Earth satellite, into an elliptical low-Earth orbit on October 4, 1957, inflicting a colossal shock. This did not happen with the Iron Dome because Israel's military research and development funds are not preallocated to different branches of the armed forces and can be spent on macroinnovations that, being entirely new, do not yet have service advocates. Instead of falling between service cracks, once their potential is recognized, macroinnovations can obtain proof of concept funding, and from then on, if a new capability is indeed demonstrated, the project can advance to production and deployment. That is the only way truly large advances can be achieved, because really new equipment is not constrained by the design limits of prior equipment and—much more important—it can benefit from a "countermeasure holiday" before adversaries can react with their technical, tactical, or operational countermeasures. It is for that reason above all that *macroinnovation* can be very important in a strategic competition: it offers new capabilities in their pristine state before they are countermeasured, and on occasion those capabilities can be decisive in war, or even in peace, that is, by forcing adversaries to redirect resources from other tasks to the countermeasures against the macroinnovation.

[2]

How Scarcity Can Force Innovation

ODDLY ENOUGH, THE FIRST MAJOR ISRAELI MACROINNOVATION WAS not anything the Israelis themselves designed or manufactured. They merely purchased the thing, but in the process, simply by specifying what components and ancillaries (i.e., subsystems) they wanted or did not want, they came up with the multirole jet fighter. Now this is almost the only kind of fighter there is, but it was in fact invented by the IDF in the late 1950s, when its Heyl Avir (Air Corps) was still insignificantly small and its resources miserably scant; it had not yet gained the prestige of its large-scale air victory of 1967.

The reason why it fell to the Israelis to invent the multirole fighter, designed to be equally capable for both air-to-air combat against other fighters or bombers and for air-to-ground delivery of bombs, rockets, and antisurface missiles, is that the leading air forces of the time—American, Soviet, British—had all come out of the Second World War with an array of aircraft highly specialized for different roles. Each category was much less capable in any other role, and they long persisted in these differentiations, sustained by ample Cold War budgets. All three air forces believed that a mix of light, medium, and heavy bombers was the best way of delivering bombloads efficiently over varied ranges against varied targets, and that air combat likewise required three different kinds of tactical aircraft: very fast interceptors that could quickly take off and fly up to engage incoming enemy bombers but had little or no ability to bomb or strafe targets on the ground; two-seater night fighters equipped with radar, not yet available for single-seaters; and escort fighters with enough range to follow and protect bomber formations. As bigger aircraft, they could also carry bombs themselves, thereby becoming fighter-bombers, but there was no special effort to develop them as such.

When Israel's Heyl Avir had its precarious start in the 1948–1949 War of Independence, the recent victory of Anglo-American airpower provided a fully proven model of what it should become, if only on a very small scale: a *balanced* air force, with interceptors, night fighters, and both light and heavy bombers, even if medium bombers were skipped.[1] That model could scarcely be applied while Israel's first and longest war was still underway because the United States and Britain refused to sell the Israelis any combat aircraft at all, forcing them to buy whatever they could find on sale. Right through the war, from May 15, 1948, until the 1949 armistice agreements, it was only by the strenuous efforts of mostly self-taught local mechanics, the priceless expertise of war-veteran volunteers from abroad, and the drastic cannibalization of a variegated inventory that flyable aircraft in twos and threes could be made ready for each day's air operations, with surges that barely reached squadron size. Nevertheless, the Heyl Avir did gradually gain air-combat superiority—ironically most clearly evident in accidental air combat on January 7, 1949, when RAF fighters on a reconnaissance patrol were engaged over the Sinai desert, with three British Spitfires and one Tempest shot down by Israeli Spitfires without loss (There was no British retaliation because this intrusion in the fighting between Israel and Egypt had not been authorized by the British government).[2]

Only after the 1949 armistice could Israeli airmen look up from the daily urgencies of combat to start building a proper air force, which would of course seek to emulate the Anglo-American model, for which bombers, fighter-bombers, fighter-interceptors, and radar-equipped night fighters would all be needed, albeit on a tiny scale, with one or at most two small squadrons of each type. In a poor country of fewer than two million inhabitants, many of them penniless refugees from Europe, North Africa, and Iraq, there were no funds to pay for such ambitions. Yet by buying odd lots of discarded aircraft, something of a balanced force with both fighters and bombers was beginning to emerge by 1953, though with very low sortie rates for want of replacement parts, when a thirty-two-year-old fighter pilot, Dan Tolkowsky, ex-RAF and with a clipped British accent, became chief of staff of the Israeli Army's Air Corps.[3]

The Air Corps of 1953 was scarcely more independent than the artillery; hence Tolkowsky could make no big decisions without the approval of his fellows on the General Staff, all infantry officers but for the equally beleaguered commander of the Sea Corps. That was a problem, because Tolkowsky flatly rejected the "balanced air force" consensus reaffirmed every day by the American, British, French, and Soviet air forces. He declared night fighters obsolete because radars would soon equip all fighters. More controversially, he believed that the Heyl Avir should have neither bombers nor fighter-interceptors, but only a single tactical aircraft for all combat roles.

At the time, the United States was developing the Delta interceptors as well as the phenomenally fast F-104 Starfighter, with no provisions at all for ground attack or bombing. The British put two engines one on top of the other to accelerate their Lightning interceptor to medium altitudes, again with no provisions for bombing, and the French—the only ones who might sell aircraft to Israel—were also designing a fighter to quickly climb high enough to launch air-to-air missiles at Soviet bombers, with no cannon, bomb racks, or underwing racks, all of which would slow down an interceptor. It is to the credit of the infantry officers who dominated Israel's General Staff that they were even willing to give a hearing to the young Tolkowsky as he contradicted the British air marshals and American generals who had so recently won the world war by arguing that air superiority could best be won by destroying enemy aircraft on the ground with an all-out surprise attack against their air bases, not by engaging in air combat piecemeal, in defensive attrition operations. That would require every combat aircraft available, and therefore all combat aircraft had to be able to carry some bombs—ruling out the interceptors prioritized by the major air forces.

In the ensuing debate, both sides tried to prove their case by citing the 1940 Battle of Britain. Tolkowsky argued that the Germans lost that contest for air superiority because they prematurely switched their offensive effort from the RAF airfields to the bombing of London. Had they kept bombing airfields, parked aircraft, personnel housing, ready rooms, and maintenance hangars, the Germans would have won. His opponents insisted that the great lesson of the Battle of Britain was that the defending RAF Spitfires and Hurricanes, rising fueled up to intercept aircraft that had to reach them from a distance, had worn down the Luftwaffe in a cumulative process—a much more reliable method, they argued, than a single all-out strike against enemy airfields that could go wrong for any number of reasons, from adverse weather to early detection that could turn a surprise attack into an enemy ambush. In the end Tolkowsky and his number two, Ezer Weizman, a much more exuberant ex-RAF fellow pilot, persuaded the General Staff to go along with their high-risk / high-payoff operational method, the all-out strike with almost nothing kept in reserve if things went wrong—the future Operation Moked of June 5, 1967.

Having won the debate, Tolkowsky and Weizman faced the hard task of actually building an air force whose combat squadrons would be equipped entirely with multirole fighters that could both fight other aircraft and deliver decent bombloads. At the time, 1953–1954, that seemed impossible because the only aircraft that could do the job was the piston-engine P-51 Mustang, an accidental wartime development that had unexpectedly proved to be exceptionally effective in both air-to-air and air-to-ground combat, but it was a fluke nobody was trying to emulate in the jet age.[4] That the United States was developing no such aircraft was of no immediate relevance because it was still US policy to sell no weapons at all to Israel, let alone combat aircraft, even as remarkably good Soviet MiG fighters started to reach Arab air forces. The British only offered jet interceptors, being especially wedded to the idea that bombing should be done by bombers—which, however, they refused to sell.[5] That left the French, whose aircraft companies lacked the prestige of American and British manufacturers, but they were very eager to sell to anyone.

The Israelis duly purchased the subsonic first-generation Ouragan fighter, the almost-supersonic Mystère, and later the supersonic Super-Mystère, albeit in small numbers. The Israelis were very glad to have them, because the Soviet Union was supplying Arab air forces with the remarkably fast and maneuverable Mikoyan and Gurevich fighters, culminating in the eternal MiG-21.[6] This still left the Heyl Avir with the unfavorable arithmetic of too-few aircraft with too-small bombloads for its ambitious all-out strike plans, because the French too preferred dedicated bombers to do their bombing and did not design their fighters to carry much ordnance in underwing racks, nor were their engines powerful enough to carry decent bombloads.[7] By the time a possible solution emerged with the prototypes of the future French Mirage III fighter in 1958, Dan Tolkowsky had retired at the ripe old age of thirty-seven (to lead Israel's advance into technology as an investment banker) to be succeeded by the irrepressible thirty-four-year-old Ezer Weizman, who long remained younger than his years.[8]

It was during Weizman's eight-year tenure that the Air Corps finally acquired the aircraft it needed: the delta-winged Mirage IIICJ, which—only because of Weizman's persistence in forcing the designers to meet Israeli requirements—became the world's first genuine multirole fighter since the accidental Mustang, agile enough for air combat against the formidable MiG-21 while also adequate for ground attack with its two 30 mm cannon and a bombload that could exceed three metric tons. Weizman's problem when he took over in July 1958 was that the eventually very suitable Mirage IIICJ did not actually exist because the French Air Force, like its American and British counterparts, wanted an interceptor with an innovative liquid-propelled booster rocket to propel it to high altitude extra fast, with air-to-air missiles instead of cannon and no bomb racks at all. Such an aircraft would have been useless for Tolkowsky's grand air strike concept. What the French aircraft did have was a good engine that allowed a speed of Mach 2.2 to be reached in October 1958—a record for Europe. It was certainly fully up to date with its area-ruled wasp-waist cross-section for supersonic flight, a good air intercept radar, all the needed avionics, and a drag chute to shorten the landing roll.[9]

What ensued was an extended debate between absurdly ill-matched antagonists. The French manufacturer Dassault badly needed the Israeli order, because by dint of saving and scraping on everything else, including its already Spartan housing, uniforms, and food, the IDF could order seventy-two aircraft all at once, a big order for what was then still a small company. But Dassault's managers and aviation engineers simply could not bring themselves to take seriously Weizman's specifications: he seemed determined to drag their revolutionary Mirage back into the past by preferring bomb racks to their splendidly innovative liquid-fuel rocket booster, and by insisting on fitting old-fashioned 30 mm cannon instead of relying on the ultramodern air-to-air missiles just becoming available. Initially the French were confident that older and wiser heads in Israel would overrule the thirty-four-year-old pilot who wanted to spoil their aircraft. Only gradually did Dassault's managers realize that the State of Israel had indeed delegated what could easily become a life-or-death decision to young Ezer Weizman, an unsettling discovery for men used to dealing with French four-star generals in their sixties, who, moreover, deferred to their expertise.[10]

In the end Weizman won out because he had the raw power of any buyer, however ill advised, and more than his share of sheer effrontery. However, it was only the sensational success of the 1967 air offensive, when the Mirages and older French aircraft were used to destroy some 400 Egyptian, Jordanian, Syrian, and Iraqi aircraft in some fifty hours of air attacks with cannons and only a few missiles, that finally vindicated the Tolkowsky-Weizman theory of how air wars should be fought and by what kind of aircraft. (Incidentally, the Dassault company earned a fortune, and their sales greatly increased around the world—though not to Israel, because the unsentimental Charles de Gaulle abruptly changed sides in June 1967, so that Israeli-specified, indeed codesigned, Mirage Vs were sold to Arab air forces, while all deliveries to Israel were halted.[11])

In the aftermath, every air force wanted what became known in English as multirole fighters, armed with cannons and not just missiles—thereby making the still very small Israeli air force a global innovation leader in tactical airpower, a role now reaffirmed by the F-35's reliance on Israel-developed air-combat technology.[12] The US air chiefs soon accepted the validity of the new multirole concept, but because their existing aircraft were all specialized—the F-104 and F-106 as pure interceptors, while the heavy F-105 fighter-bomber lacked maneuverability for air combat, faring poorly against MiG-21s over Vietnam—the USAF was forced to acquire the Navy's versatile F-4 Phantom to serve as its first multirole fighter since the Mustang, retired long before. The humiliation was severe, and the next USAF fighter, the F-16—which would remain in production for decades—was inspired by the success of Ezer Weizman's version of the Mirage III. It was a remarkable sequence: because the structure of the IDF favored innovation—not least because their two-careers model kept even their generals young—and because their small air force could not hope to prevail without aircraft that did not yet exist, their young chief used his momentary commercial power to browbeat a European aircraft manufacturer into producing an aircraft that became the global model of what a tactical aircraft should be, even though the pre-1967 Heyl Avir was still an insignificant presence in world aviation.

[3]
A Youthful Officer Corps

AN IMPORTANT FACTOR IN THE MULTIROLE FIGHTER INNOVATION, AS in many other IDF innovations, is so elemental that its deeper significance is easily missed: the two successive Israeli air chiefs who were the innovators of their generation were still in their early thirties when they assumed command—that is, at least twenty years younger than their youngest counterparts in other air forces. Military establishments that bear the burden of life-and-death responsibilities cannot be free and easy in taking risks, and are therefore inclined to rely on well-proven personnel and well-proven methods. Military career paths are therefore meant to ensure that officers accumulate professional experience commensurate to the scope of their responsibilities, as they are promoted rank by rank. In the process, however, officers acquire wisdom and prudence in place of the recklessness of youth and a proper respect for the sound practices of the past as opposed to the uncertain advantages of untested innovations. As people become older they typically become more set in their ways, less courageous, less inclined to try out what is new, less likely to view the existing order of things as imperfect and capable of improvement, less open to the very first step of any process of innovation—the willingness to take a risk on the new in the hope that it might be much better than the old.

In the formative years of the IDF in the 1950s and early 1960s almost all senior commanders were in their thirties. In 1956, on the eve of the Sinai

Campaign, the most senior officer in the field, area commander Major General Asaf Simchoni, was thirty-four; lead brigade commander Ariel ("Arik") Sharon was twenty-eight; and IDF chief of staff Lieutenant General Moshe Dayan was forty-one. Even high-ranking positions were staffed by young and relatively inexperienced officers still learning the ropes. When Dayan was himself still an inexperienced chief of operations, he asked Aharon Yariv, a very bright, even younger officer (and future military intelligence chief), to establish and command a national defense college for Israel, such as the ones he had visited in France and England. Yariv replied that he lacked the necessary knowledge and experience. But Israel was then a newly formed state and most of its leading positions were staffed by inexperienced people. Dayan answered, "If Yitzhak Ben Zvi can be president, Moshe Sharet Prime Minister, Maklef can be Chief of Staff, and I can be head of the Operations Directorate—you can be the commander of the Defense College."[1]

The simple chronological explanation for this state of affairs—that there could be no experienced veterans in a very young state—turned out to be wrong in the case of the IDF. The armed forces of the new Israeli state born in 1948 started off with officers in their twenties, who reached their thirties in the 1950s, as Tolkowsky and Weizman did. But then the progression stopped, for otherwise the mere passage of time would have yielded a normal age structure by the later 1970s, with admirals and generals in their early sixties (as in the US and European armed forces), colonels in their late fifties, and so on. It was Moshe Dayan, chief of staff from 1953 to 1958, who instituted a dual-careers principle whereby the IDF would give midcareer leave to officers to pursue their education (given that hardly any had progressed beyond secondary school), which would enable them to embark on a second, civilian career after an early retirement in their thirties. Dayan explicitly stated the goal: he wanted to keep the officer corps mentally young and flexible, physically agile, and with the bravery of youth.[2]

Dayan acted accordingly, retiring as chief of staff at age forty-three, but he did not try to turn his policy preference into a permanent practice anchored in a law. It turned out, however, that a precedent had been set, because the seemingly inevitable progression toward a normal age structure in the IDF underway since 1948 (when the highest field commander, Yigal Allon, was thirty) did not continue. Chief of the General Staff of the Israel Defense Forces Lieutenant General Aviv Kochavi took office on January 15, 2019, at age fifty-four, when he was the oldest member of the General Staff, commanding an officer corps whose generals are some ten years younger than their American and European counterparts. His predecessor as chief of staff, Gadi Eisenkot, had inaugurated his tenure by calling for a younger officer corps, and successfully lowered the average retirement age from forty-seven to forty-two.[3]

The biological reality of relative youth seems to make a difference, diminishing well-aged wisdom no doubt, sometimes with serious consequences, but certainly allowing greater scope for innovation. It is true of course that youthful exuberance and optimism ensure nothing more than a willingness to consider the new, and in no way guarantees the mental ability to invent it—but even a mere tolerance of the new and untried is in scarce supply in most military establishments. Even research and development departments, whose very mission is supposedly the advancement of innovation, are often bastions of tradition, as they keep offering updated versions of established platform and weapon configurations, which they resolutely defend from anything really new that comes along. Older officers, whose future is shorter than their past, are much more likely to demand the perpetuation of the iconic platforms and weapons of their particular branch and their own service, thereby denying funds for new things—even while joining the call for more innovation everywhere else.

Nor is the great lesson of *macroinnovation* ever learned: only if they are truly new, not just new and improved, can new weapons or techniques secure the great reward of a "countermeasure holiday" from the counterweapons and countering tactics evoked by their predecessors.[4] An entire generation of clever Englishmen had a good laugh when Field Marshal Douglas Haig explained in 1925 that while airplanes and tanks had their uses, "they were only accessories to the man and the horse," adding that he was "sure that as time went on, [soldiers] would find just as much use for the horse—the well-bred horse—as they had ever done in the past."[5] In commanding the British Army on the Western Front from 1915, Haig had presided over the useless immobility of men and horses (however well bred) brought about by the machine gun, and then over the victory of the first tanks that could ignore machine-gun bullets as they advanced. Hence his vision of a future for the horse in battle was a simple case of temporal dissonance: he said in 1925 what might have been plausible in 1900 but had already been invalidated in 1904 by the Russian machine guns and barbed wire at Port Arthur during the Russo-Japanese War.

That is easy to see now, almost a century later, while today's examples of temporal dissonance are much less obvious. See, for example, the case of the US Air Force's future B-21 Raider long-range (i.e., strategic) nonnuclear and nuclear bomber, which, as noted earlier, is to be manned by a crew and also optionally unmanned, thereby adding all the robotic costs to the man-rating costs and the return-home cost, a very heavy price indeed to allow a few air force officers to pilot those aircraft. In this case the temporal dissonance achieves Haig-like levels because as of 2021, the B-21 was still in its engineering development phase as a manned bomber, even as unmanned, self-driving trucks were already operating experimentally on Arizona highways, in a far more complicated environment than the untextured skies.

Temporal dissonance has been a chronic problem for armed forces everywhere in this age of revolutionary technological change, which arguably started on September 26, 1825, with the inaugural run of George Stephenson's steam-powered Stockton and Darlington Railway. The IDF's answer for this problem is to rely on youth, for whom the past is much less important than the future, and is therefore less invested culturally and intellectually in the iconic weapons and concepts of the past still in service, whose perpetuation precludes the emergence of the new.

A Deliberate Shortage of Senior Officers

Youth is important in officers, but so is scarcity. In war, there might be a great need of an abundance of junior officers in the combat arms, especially infantry, in which junior officers are akin to ammunition and may be expended as such, in order to advance against enemy fire. But there is rarely a shortage of senior officers; indeed there are too often too many of them—though not in the IDF, whose officer corps is as notable for its extremely compressed rank structure as for its youth.

Instead of the rank inflation often deplored elsewhere—NATO gatherings feature entire regiments of three- and four-star admirals and generals, and it was an abundance of stars that prompted the US Congress to enact a legal limit—the exact opposite is true of the IDF, in which large responsibilities are often assigned to relatively low-ranking officers because of the chronic shortage of senior officers.[6] Given that the country's top military officer, the chief of staff (*ramatkal*), is himself a three-star lieutenant-general (*rav aluf*), with no four-star generals or admirals above him, a downward compression of ranks follows inevitably—and it is certainly extreme. The entire Israeli Air Force is commanded by a two-star major general, even though it now operates more aircraft than the British Royal Air Force, which is commanded by a four-star air chief marshal, with several three- and two-star air marshals below him.

The same compression of higher ranks is present but less glaring in Israel's navy, Ḥeyl HaYam, which is still a small force by global standards, unlike the air force, yet is no longer a miniature navy either, with a significant submarine flotilla as well as four corvettes, eight large missile boats, and numerous patrol boats. Its commander is also a two-star officer, whose Hebrew rank of *aluf* is translated as rear admiral, while many of his peers around the world in charge of less capable navies are three-star vice admirals or more often four-star full admirals.[7]

Besides the air and naval chiefs, there are only seventeen other IDF major generals (*alufim*) on active duty. HaKirya ("The City"), the IDF general headquarters in Tel Aviv—Israel's Pentagon, more or less—is headed by just nine major generals, a fraction of the number employed in the general headquarters of any comparable armed force elsewhere.[8] That short list begins with the deputy chief of the general staff, who is not supposed to be in the same room as the chief when the threat level is extreme, and who functions as the actual chief of staff when any combat is imminent or underway. The top officer, though called chief of staff, actually functions as the commander in chief, in direct charge of all ground, naval, and air forces (in contrast to US practice), mostly leaving it to his deputy to do the coordinating chief of staff function.[9]

Next in the chain of command is the ground counterpart of the air and naval chiefs, the head of the "ground forces command," Mifkedet Zro'a ha-Yabasha, who overviews their development in peacetime, and the allocation of the field units between battle fronts in wartime. In other words, the top operational command of the entire IDF is the responsibility of just five officers, even though the IDF become among the world's larger military forces, numbering some 650,000, when fully mobilized.

Like all their modern counterparts, the IDF have different general staff departments, which, as everywhere else including China, still follow the classic nineteenth-century Prussian model attributed to Field Marshal Helmuth Karl

Bernhard Graf von Moltke (chief of the German General Staff from 1871 to 1888), with some twenty-first-century additions. The operations department, Agaf Mivtza'im (or G-3), coordinates (but does not command) campaigns in wartime, and strives to plan them in rough outline in peacetime; military intelligence, Agaf HaModi'in (or G-2), whose acronym AMAN is the subject of many romantic tales, is mostly dedicated to analytical desk work. Then comes the post-Prussian technology and logistics directorate, Agaf ha'Technologia ve ha'Logistica (or G-4 department), successor of the old Prussian quartermaster department, which supplies everything including the army's notorious canned meatloaf (LUF, aka "donkey meat" rations); the personnel directorate, Agaf Koach Adam (to call it a "manpower" department would be an especially poor translation given that Israel's army has the most women of any in history); the new planning and force development directorate, Agaf ha'Tichnun, which is supposed to predict the future (a local tradition to be sure) and devise ways in which the IDF can adapt to cope with future needs, both much harder and much easier back in the days when Israel was still threatened by large conventional armies; and the head of the twenty-first-century computer service directorate, Agaf ha'Tikshuv, whose importance has been increasing in almost linear fashion since the IDF acquired their first IBM 360 mainframe computer soon after its 1964 appearance. Lastly, there is the new addition of the Strategy and Third Ring directorate, Agaf Estrategia Vma'agal Shlishi, established in 2020 to focus on the Iranian threat.[10]

Beyond the general headquarters, three more major generals, each in their own regional headquarters, are in charge of the area commands for the North (*Pikud Tzafon*), Center (*Pikud Merkaz*), and South (*Pikud Darom*). It is their responsibility to overview the frontal perimeters, with their fences and patrols and all the supply and mobilization depots, where reservists would arrive in the thousands upon mobilization to find uniforms, weapons, transport, and heavy weapons, including tanks supposed to be ready and waiting to roll out into battle. In wartime the territorial commanders become front commanders if there is large-scale fighting, responsible for deploying the forces assigned to them by the general staff and coordinating them in action.

With all three area commanders focused on military operations beyond Israel's borders, while internal threats abound, another major general heads a Home Front command (*Pikud Ha'Oref*). Also, though never very willingly, the army must operate in the mostly Palestinian-inhabited West Bank, so another major general has the carefully worded title of "Coordinator of Government Activities in the Territories" (*Me'ta'em ha'Pe'ulot ba'Shtakhim*).

All these positions are filled by just sixteen two-star major generals and that includes the defense attaché in Washington, DC, who also enjoys the luxury of a proper dress uniform, unlike his colleagues at home who must make do with field uniforms, direct descendants of the battle dress of the British armed forces of the Second World War, with none of the gold braid of their foreign counterparts.[11] Any comparison of Israel's lone lieutenant general and small band of major generals with the 144 two-star, 68 three-star, and 39 four-star flag-rank generals and admirals of the US armed forces would be meaningless, given the latter's sevenfold larger size and, even more important, the global scope of their functions in both theater and alliance commands. Emulating the British formula in fighting the formidable Napoleon, of recruiting all possible allies so as to fight the lion with as many dogs and cats as possible and a few mice thrown in, American grand strategy requires the upkeep of as many alliances as possible, which necessitates a very patient and constant military diplomacy that is mostly the responsibility of flag-rank officers. Still, Israel's single three-star and twenty-four two-star generals may be validly compared to the hundreds of generals that populate Europe's shrinking armed forces, greatly outnumbering their combat formations, warships, or aircraft.[12]

The reason why the numbers at the top are important is that scarcity at the top necessarily drives responsibility down the chain of command, with very positive effects when it comes to innovation. How that works in the IDF is far from straightforward, but one aspect is simple enough: because the generals themselves in their forties are too few, innovation decisions are mostly made by subordinates in their thirties who are much less shaped by the past and much more open to the future.

Forcing Responsibility Downward

The compression of ranks that starts at the top with only one three-star lieutenant general for the entire IDF, and very few two-star major generals, continues down the line. In modern armies around the world, three brigades form a division, with 10,000–20,000 troops under a two-star major-general, but in Israel divisional command in the active-duty forces only rates a one-star *tat aluf,* officially translated as brigadier general. Of the latter there are roughly seventy-five for the entire IDF, including their small navy and large air force, which has many more combat aircraft in operating condition than any European air force.

Because the IDF have neither the global deployment structure of the US armed forces nor the elaborate bureaucratic overhead or inherited historic facilities of the European armed forces, IDF ground forces have a very high "teeth-to-tail" ratio, with many combat formations for its total size: some thirty-six active-duty and reserve brigades of 2,000–3,000 soldiers each in the ground forces (more than in any current NATO army). Those brigades are commanded by colonels, as in most armies, but they are distinctly younger colonels, some under forty. The three battalions of a typical brigade are each commanded by a lieutenant colonel, as in most armies, but again they are much younger, with an average age around thirty. It follows that the three companies of a typical active-duty battalion are each commanded by a captain, again as in most armies, but at twenty-five or so, their average age is lower than in other armies.

The inevitable effect of a very small number of senior officers is to thrust responsibility downward, initiating a cascade effect as field-grade colonels, lieutenant colonels, or even majors find themselves charged with duties and responsibilities that generals cannot do because they are already fully engaged in even more essential tasks. And because the field-grade officers are also too few by any normal standard (the total number of colonels in the entire IDF is roughly 450 for a force that exceeds 600,000 at full mobilization), many of their responsibilities fall to more junior officers, so that captains in their twenties are routinely entrusted with duties that are elsewhere reserved for older, decidedly more senior officers.

The scarcity of senior officers, the resulting compression of ranks, and the thrusting of responsibilities downward finally meet the contradiction of a great abundance of junior officers, because they are not the scarce, expensive products of multiyear military academies as in the United States and other countries, but rather young conscripts who have signed up to serve one more year in order to apply for and if possible attend the IDF's demanding officer course, a socially important attainment. Having started as eighteen-year-old recruits mobilized for three years of compulsory military service, junior officer candidates are selected from the ranks to attend the officer training course. That requires a further year of service after the course, for a total of four and half years in uniform; hence not all who are offered the opportunity to become officers choose to accept.

This is when another peculiarity of the IDF's officer cadre intervenes: instead of the overabundance of midranking officers often deplored elsewhere, the IDF has very few of them, in fact absurdly few by global standards. Israeli lieutenants, only a year or two removed from boyhood or girlhood, can find themselves responsible for the lives of many others, commanding thirty-person platoons, or several tanks, or artillery batteries with their crews, or patrol boats in the navy, unless they serve as flying officers in the air force, habitually in charge of vastly expensive, devastatingly powerful aircraft. That is true of any armed force of course—in all of them young lieutenants must shoulder vast responsibilities if there is actual combat. But where the IDF differs, and it is a very large difference, is in the absence of professional noncommissioned officers. NCOs are the normally older, typically much more experienced sergeants and warrant officers considered the backbone of the US, British, Russian, Chinese, Indian, and indeed almost all armies—except the IDF. With NCOs around, young officers are never alone.

The IDF has no such cadre, because the divide between officers and NCOs in more established armies is the product of class differences that are meaningless in the Israeli context, where there are wide income gaps but not the corresponding social distance between gentlemen who might be officers and able but lower-class youths content to be sergeants.[13] In any case the IDF has never had a military academy to turn school-leavers into officers and gentlemen, whose graduates are socially superior to their noncommissioned inferiors, who remain their subordinates even if more experienced and more skilled. Hence even very young IDF officers are left to deal with both dangerous and also delicate situations entirely on their own, with no superior officer or experienced NCO present to guide them in person. Inevitably, these young people are much better suited to deal with the dangerous rather than the delicate.[14]

A typical track for future infantry officers starts with twenty-eight to thirty weeks of basic training given to all recruits, followed by twelve weeks of specific infantry training. Next is an assignment to a specific battalion for another twelve weeks of routine security and combat training. Only then is a recruit considered an infantry soldier capable of combat, having absorbed a cumulative total of about forty weeks of fieldcraft, weapon, and tactical training, twenty-four of them intensive. That is not much, one might think, to turn an eighteen-year-old young man or woman into a soldier who might be in combat at any time, but actually it comprises more training than the US Army or Marine Corps provides to enlisted infantry, whose combined basic and advanced infantry training may only amount to twenty-three weeks, or even less; in the US Army it is only recently that initial infantry training has been extended from fourteen weeks to twenty-two.[15]

IDF soldiers who do well as recruits through their first year or so are selected for the fourteen-week squad leader *kurs makim* course, the most important tradition inherited from the prestate Haganah, for which it amounted to just about the totality of training. Along with a strong emphasis on the duties and skills of junior combat leadership—heady stuff but also sobering for nineteen-to-twenty-year-olds—the *kurs makim* adds another layer of tactical training. And those weeks are more valuable than the previous training weeks because of the higher caliber and more intense motivation of the average trainee. All who successfully pass out of the *kurs makim* return to serve as squad leaders for at least three months. Some remain as sergeants in their battalions until the end of their conscript service, while others are diverted to specializations of one sort or another, but the best are offered the opportunity of becoming officers by attending the thirty-week officer training course, the *Ba'had Ehad* of many youngsters' aspirations.

As already noted, the US and British armies rely heavily on experienced noncommissioned officers, not least to guide newly baked second lieutenants who might be twenty years younger. The IDF's lack of such NCOs (its own are nineteen to twenty-one years old) means that young officers posted to commands must assume full responsibility from day one—it is only when reserve units are mobilized for special exercises or for war that the IDF acquires distinctly older, battle-hardened NCOs, because to serve fifteen or twenty years in the same reserve company is a common practice. But normally, without older reservists to help them, the new products of the officer training course must do it all, and that is a great deal, because those young officers must contend with the large responsibilities that Israel's understaffed command structure relentlessly delegates downward.

Ba'had Ehad's standards are therefore jealously guarded against every contrary wind that would reduce its admission criteria and performance requirements, from antielitism to worries about the impact of required desert marches in maximum heat for the products of air-conditioned childhoods. Standards are kept high yet the failure rate is still low because admission is seriously selective, including in-depth psychological testing, which has always been held in high regard by the IDF, whose psychology unit itself attracts highly talented conscripts. (The future economics Nobel Prize winner Daniel Kahneman contributed a practical test to the selection process in 1955, when he was a twenty-one-year-old "psychology officer").[16]

Upon graduation, new second lieutenants are appointed to lead a platoon. By then they will have accumulated almost two years of experience as soldiers and junior leaders, with much intensive training. If all goes well, promotion to lieutenant follows routinely, and the required one year of additional service—with a modest salary, as opposed to the mere pocket money given to conscripts—can become the start of a military career.[17] Outwardly, these young Israelis resemble their American and European counterparts, certainly in their basic duties: to command soldiers in security surveillance missions and in combat; and to overview their training in peacetime. As with other armed forces around the world that can actually fight armed enemies, and not just strut about in uniform, for IDF junior officers combat leadership is the essence of the profession. And that means leading from the front; not "advance" but "follow me" is the classic Israeli command, as in other fighting armies.

But that is where the similarities end. First, as noted earlier, the IDF have no military academies—there is no West Point or Annapolis, no Royal Military Academy (Sandhurst) or École spéciale militaire de Saint-Cyr, where graduates or near enough are turned into officers without ever having served as enlisted men, after two, three, or four years of mostly academic education. The very existence of military academies reflects class distinctions that are in great part outdated; it was once felt that higher-class youths could not be asked to train and live among uncouth troopers, who were mostly uneducated or even illiterate, unfit for the literate aspects of military education.

The classic military academies produce many of the future senior officers of the US, British, and French armed forces, but officer candidate schools open to both college graduates and selected noncommissioned officers have historically supplied the majority of officers. Dispensing with the prolonged if largely civilian education offered by the military academies, they impart accelerated training over a few months. The US Army's version, advertised as "a rigorous 12-week course designed to train, assess, evaluate, and develop second lieutenants for the U.S. Army's sixteen basic branches," is offered to noncommissioned officers as well as to college graduates who have completed a ten-week basic combat training course, thus producing second lieutenants who may be sent to command a platoon in war in a total of twenty-two weeks from the day they joined the army.[18]

The IDF's lack of military academies—their explicit elitism would have been totally out of place in a socially egalitarian society—along with their practice of promoting officers from the ranks with no university education or a prior noncommissioned career, results in a paradox. The IDF's junior officers are much younger than their foreign counterparts and much less educated, but they receive much more military instruction before taking up their first command—more than three times as much as many US Army officers. That turns out to make a big difference in modern war, which typically lacks both the horrors of high-intensity warfare with its artillery barrages, and its on-the-job training. As for endless counterinsurgency wars, they are if anything diseducational, because those who learn war in Iraq, Afghanistan, or the West Bank are not learning how to fight well-armed enemies but only acquiring bad habits in fighting irremediably weak enemies with no armor, no artillery, no airpower, and no overhead intelligence.

For IDF aviators, the absence of an air academy, and their start in uniform at age eighteen instead of a postcollege twenty-two or older, results not merely in differences of one sort or another, but rather in a complete reversal: in the US Air Force, as in the British and all European air forces, future pilots are first educated to be officers before they start flying combat aircraft, while their Israeli counterparts first serve as pilots before they are educated to become officers, if they so desire. That is the case for both the four-year US Air Force Academy at Colorado Springs and the Royal Air Force College Cranwell, whose thirty-two-week course more or less presumes prior university education, given that its subjects include "transformational leadership and academic air power studies, including ethics and strategic thinking as well as more prosaic military skills, essential service knowledge, drill and physical training."[19] Flight training only starts after that, in the twenty-one-month "fast jet" course, for example, which begins with flights in a light aircraft and then a powerful turboprop before progressing to a jet trainer, and then onto a Royal Air Force Operational Conversion Unit, in which pilots are finally trained to fly and fight a front-line fighter at age twenty-five or so.

Their Israeli counterparts, by contrast, are flying light aircraft within six months of enrollment, while still nineteen years old at most, so that their elemental piloting talents, or lack of them, can quickly be determined. Those who pass muster and accept the obligation of additional years of military service are trained for two years in a variety of skills (including infantry combat) and also educated in a variety of academic subjects, while continuing their flight training. In their third year in uniform at age twenty-one or twenty-two (when their US and European counterparts are just starting) they are fully trained pilots assigned to operational squadrons of jet fighters, transports, reconnaissance, or other aircraft, including helicopters.[20]

It is a system that has produced many excellent young pilots who have shot down many enemy aircraft and scored many a bull's-eye in ground strikes, but who are not sufficiently educated to manage a large and technologically ambitious air force. That is the task of the relatively few career officers who remain in the IDF after completing their compulsory service with the aviators' extra years, and who are given generous study leave at full pay to acquire the necessary higher education. Given the scarcity of senior officers, middle-ranking officers must necessarily assume their responsibilities; in the US Air Force, a three-squadron air wing might be commanded by a forty-year-old colonel or one-star brigadier general, but in Israel it would be an officer some ten years younger of much lower rank.

[4]

Innovation from Below

BECAUSE OF THE CHRONIC SHORTAGE OF SENIOR OFFICERS, IDF junior officers, in spite of their young years and thin education, are routinely compelled to assume disproportionate responsibilities. Faced with complicated situations, when circumstances deny immediate guidance from older officers on the scene such as recalled reservists who happen to be there, junior officers must take, or better seize, the initiative by devising and executing their own schemes of action to avert sudden dangers or exploit fleeting opportunities. That requires the power of decision, and perhaps bold leadership, but such qualities only become relevant when the young officer has devised a plan of action in a situation that was perhaps never anticipated during training or in the guidance, if any, that came with the task.

It is true of course that any protracted conflict will feature recurring threats and recurring opportunities, which indeed become the subjects of training courses, even of prior planning.

But such is the crooked timber of humanity, and such is the infinite variety of circumstances in which any episode of conflict may unfold, that no prior analytical process can anticipate all the variables and uncertainties that will determine the best course of action in any given situation.

Therefore, in all modern military forces young officers are constantly enjoined to seize the initiative, in order to respond as advantageously as possible to unpredictable circumstances by formulating and executing a plan of action before the situation changes. In other words, even though every junior or midranking officer exists within a hierarchical chain of command populated by more senior officers to whom obedience is due, any officer must be mentally prepared to think and act entirely on his or her own to react quickly enough to seize fleeting opportunities and avoid sudden dangers. Hence young US officers in training are told to prepare themselves to seize the initiative, and they hear the same from more senior officers upon their first commands.

There is an excellent reason for all this insistent preaching: in most forms of warfare, nothing is more power enhancing than a superior propensity to take the initiative—it can overcome even gross inferiorities in numbers and firepower.[1] It is a matter of relative velocities in action and reaction. Just as a fleet-footed boxer can knock out a much stronger opponent who keeps missing him with more powerful punches, an armed force whose officers down to the junior ranks are able and willing to take the initiative is much more agile. It can move, act, and react more swiftly, landing its own punches while sidestepping those of the enemy. On a larger scale, that agility makes an armed force more capable of *maneuver* warfare, in which the aim is to minimize casualties while maximizing gains by deliberately circumventing enemy strengths and boring down to exploit detected enemy weaknesses. That is opposed to the far more common *attrition* warfare, in which strength is frontally activated against enemy strength, in grinding-down forms of combat in which victory goes to the side that can better withstand material losses and human casualties.

Because no enemy will willingly stand by while its strengths are circumvented and its weaknesses are exploited, the contest is decided by the relative velocity of each side's action, which in turn will depend—other things being equal—on the relative propensities of officers, up and down the chain of command, to take the initiative and act. More battles, indeed entire campaigns, have been decided by invisible and unmeasurable initiative imbalances than by material imbalances, as is certainly the case with Israel's battles and campaigns.

Yet to teach, promote, encourage, and even demand the initiative is often useless; it may be counterproductive in armed forces whose officers are just not up to it, who are not willing to risk all on their own judgment of the situation. They will do nothing while awaiting orders because of fear of failure, thereby becoming useless as decision-makers, whereas they could have been even somewhat useful if sent orders from above that were simply to be obeyed. The reason is simple: the actual extent of the initiative that can truly be exercised by real-life officers in real-life armed forces depends on their *structure* above all. That is, it depends on what they really are, as opposed to what they say they are or what they would like to be, specifically in their command-and-control practices.

Nobody exceeds the US Army, Navy, Air Force, and Marines in stressing the importance of taking the initiative, especially in staff and command courses where field-grade officers are trained. At the same time, however, no armed forces in the world have better surveillance systems with which more senior officers can monitor their subordinates, and no armed forces in the world have better telecommunications with which more senior officers can send instructions to subordinates—indeed they are called command-and-*control* systems. Finally, the US armed forces are the most amply organized and structured, with the most elaborate headquarters, and the largest staffs at each echelon. For example, the G-3 or Operations section of a US Army divisional headquarters might have as many as twenty officers, as opposed to three or four in a German division, when they still existed in the 1980s.

Therein lies the problem: US officers are not idle fellows, content to draw their pay without exerting themselves overmuch. They have a very strong work ethic: if they are assigned to serve as staff officers in a divisional G-3, one can be sure they will work hard to generate an unending flow of detailed orders, warnings, and redirections for the brigade commands below them, within each of which is an equally hard-working G-3 officer who does the same for the battalions of that brigade. In spite of all the good teaching and exhortations younger officers hear about taking the initiative, about sizing up a situation and acting swiftly instead of just reporting back and awaiting orders, the inevitable result of all those officers above them so well equipped to monitor them is to reduce the perceived freedom of action of more junior commanders, and thus their propensity to think up initiatives and execute them on their own responsibility.[2]

It is useless to exhort officers to act boldly in devising and executing their own self-initiated plans when so little room is left by the constant interference of higher command echelons.[3] In other words, even in the absence of the culturally mandated rigidity of many armies, whereby junior officers are always on standby awaiting orders, this matter of the initiative, which is really of the greatest importance, does not depend on what is taught in military schools. Rather it depends on the very nature of the command echelons at each level: thick ones with lots of officers restrict initiative downward while thin ones with few officers force responsibility downward.

In the IDF seizing the initiative is not a matter of choice: understaffed command posts with understaffed headquarters above them can only issue broad, undetailed mission orders to subordinate commanders. Simply, they can only define *what* needs to be done: seize this, hold that, clear an area, adding a warning perhaps (land mines!), but they give no instructions on *how* the thing is to be done. That is left to the commander on the spot, who may be a very experienced sixty-year-old retired general, recalled to command an entire reserve division mobilized for war, or a twenty-one-year-old lieutenant in charge of a platoon.

What is missing is exactly what keeps headquarters staff busy in overofficered armies: the issuance of detailed orders for subordinate units that specify what is to be accomplished and *how* it is to be accomplished, perhaps with lots of tactical specifics, prescribed routes, recommended modes of action, detailed fire-support plans, supply provisions, and more, as opposed to the few words of a typical mission order.[4] The front commander Yigal Alon's written order for the first-ever large-scale IDF operation (Operation Yoav in October 1948), an all-out counteroffensive to drive back the invading Egyptian Army, consisted of a single page.

Unit commanders who receive such detailed directives but then encounter an unexpected obstacle when trying to implement them—which happens all the time in war because enemies must strive to block whatever is underway—cannot simply improvise their way around that obstacle. They must refer back to the headquarters that issued those detailed orders, describe the unforeseen obstacle, and ask for new orders. While awaiting word from above, the unit must pause. The overall effort thus becomes a series of interrupted actions—and each interruption provides breathing room to the enemy, allowing it to devise and execute countermoves. Soldiers serving in such armies quickly become used to a go, stop, go-again, stop-again sequence while their immediate commanders report up the chain of command and await orders at each remove. (This was true of the US forces that landed in Italy and France in 1943 and 1944 to drive back the Germans, except for the forces under George S. Patton's command.)

With mission orders—and the entire risk-taking mentality that goes with them—initiatives at each level of command replace directives from above. Unit commanders can command their own action, immediately responding to unexpected obstacles by maneuvering around them or taking whatever other action might be appropriate, without referring back to headquarters or stopping to await new orders. Powered by the unencumbered exercise of the initiative by commanders down the chain of command, the overall action can be fast and fluid, with obstacles circumvented in a continuous sequence of moves, even if there are a lot of zigzags as unit commanders try to find the best way forward and try something else if they encounter too much resistance. If the enemy happens to have a strictly top-down command system (as is true of most armies around the world), commanders at each echelon will be tied down waiting for new orders, still in the process of reacting to the previous move even as a new move is unfolding.

That asymmetry was certainly evident during the thirty years of Arab-Israeli conventional warfare that ended with the 1973 cease-fire, but for a few days of major combat against Syrian forces in Lebanon in June 1982. After that, with no further wars of movement, the IDF became absorbed by repetitive security operations, in which there was little room for the exercise of initiative. But then came the so-called Second Lebanon War of July 12–August 14, 2006, followed by an official commission of inquiry whose brief was to evaluate every aspect of the IDF's performance, as well as the government's decision-making process.[5] The final report, publicly issued on January 30, 2008, was harshly critical, especially regarding what it described as "an overcentralized style of command" said to have inhibited the initiative of subordinates.

Foreign observers, already surprised by the complete indifference to prestige or even security considerations in exposing everything that went wrong in the course of winning a war, were puzzled by the overcentralization finding regarding practices that are perfectly normal in other first-class armies.[6] But in Israel itself the criticism resonated strongly, because a great many Israelis had combat experience and some expertise in military matters and were therefore rightly concerned by any sign of diminished freedom of action for subordinates down the chain of command, an unwanted consequence of technological progress in telecommunications.

The Initiative as the Path to Innovation

There is a direct relationship between the IDF's culture of initiative in combat and technological innovation. Officers who learn to take the initiative, whose minds are formed by the mission-order mentality, are much more likely to view critically what is around them, be it army tactics, operational methods, procedures, the equipment issued to units, or even the cut of uniforms. And they are much less likely to accept limitations and shortcomings as inevitable. Instead of learning to accept, or even cherish, inherited organizational formats, specific weapons issued to their units, prescribed tactics, and standard operating procedures, they tend to question everything—and then strive to come up with their own supposedly better answers. That is the larval stage of all innovation: the increasingly detailed questioning of what there is, preparatory to formulating alternatives.

It is no wonder that many young IDF officers proceed straight from military service into start-up ventures with fellow soldiers, and not only in the high-tech sector.[7] That is true not only of those who served in one of the IDF's more technological units but also of those who served in ordinary combat units. In a fairly typical experience, when former Jerusalem mayor and Knesset member Nir Barkat completed his military service as a captain in the Thirty-Fifth Paratroopers Brigade, he founded a software company (BRM) that successfully developed new antivirus software for the internet. Years later, he described his transition from the army to the startup world: "The feeling around us was that we were part of an elite group and that the sky was the limit . . . therefore, [one is] allowed to make some mistakes . . . I felt immediately at home . . . there was room for bold actions, trial and error, just like the paratroop unit I served in."[8]

In September 1968, author Edward Luttwak visited the office of MG Mattityahu ("Matti") Peled, then head of the General Staff's quartermaster department (now the Technological and Logistics Directorate) responsible for the supply of everything from boots to missiles, and subsequently a distinguished professor of Arab literature and a peace campaigner. Peled had just received the first sample of a new combat harness for the infantry—an open-mesh nylon vest that seemed very practical with its four rifle-magazine pockets, four hooks for hand grenades, a pocket for the first-aid kit, and a lightweight fiber belt with fittings for two water bottles. The harness and belt were obviously a huge improvement over the existing webbing, magazine holders, grenade pouches, and belt that were ex-British surplus Second World War gear made of very stiff, heavy woven cotton, with awkward buttons and ill-fitting straps.

When asked if he had solicited suggestions before starting the design process—after all, a high proportion of all Israelis would be wearing that webbing—Peled burst out in good-natured laughter and said he had not issued any announcement nor solicited any suggestions. But word had got out anyway; every single infantryman who had ever served in the army already had his own far more advanced design, at least in his own mind. And what was the reaction when news of the chosen design started circulating? Oh, just the usual reaction, Peled said: nobody had actually seen it, but everybody already knew that the new harness was a complete failure, what with its flammable nylon, flimsy open mesh, and loose hooks instead of pouches for the grenades. As it was, the critics were wrong, and the new combat harness turned out to be a great success—except that the grenade hooks were replaced with pouches—because it already embodied creative responses to all the uncomfortable years of marching and fighting with the old British webbing.

Nevertheless, it is obvious that however lively their minds, the young conscripts who make up the bulk of the standing army, along with the ex-conscript lieutenants serving an extra "professional" year or two and the small cadre of career professional officers, are not exactly well placed to come up with important innovations. Conscripts are high-school graduates at best, and at the ripe old age of eighteen cannot have accumulated much prearmy work experience either. It is true of course that their technological creativity may be greatly stimulated by all the high-tech gear they operate in their army service, but their lack of a university education must limit their ability to evaluate technological innovations with proper quantitative methods. As for career officers, who do have a university education, they are both few and badly overworked, reducing their ability to pursue innovation, which is hardly their mission.

It does help that Israel is a largely informal society in which there is a general readiness to talk to anyone and even listen to them up to a point. Nor is it hard to reach IDF decision-makers and have them listen to any reasonable new idea. Edward Luttwak was not a reservist but a newcomer to Israel in the summer of 1970, when the Israeli Air Force was flying over Egyptian territory across the Suez Canal to destroy the echeloned antiaircraft missile batteries the Soviet Union was generously supplying. The batteries were being emplaced closer and closer to the Suez Canal front line, diminishing the ability of Israeli fighter-bombers to strike back against the Egyptian artillery batteries constantly bombarding (thousands to tens of thousands of shells per day) the vastly outnumbered Israeli troops and against an eventual Egyptian canal-crossing offensive. Opening the skies for effective aerial operations in support of the ground troops required destroying the missile batteries, and this in turn was predicated on locating them and conducting carefully planned and precisely conducted strikes. Locating the Egyptian artillery and surface-to-air missile batteries in order to strike them successfully required aerial photography that had to be conducted at medium altitudes with aircraft flying a straight path at a constant speed—a perfect target for missiles. Evading the missiles required aborting the photographic mission to conduct violent aerobatics. As a result, though only one Israeli photo-reconnaissance aircraft was destroyed by a missile, many missions were aborted, reducing the effectiveness of the air force's support essential for the ground forces.

With zero credentials, Luttwak reacted to the distressing news by seeking an appointment with Aryeh Dvoretzky, a leading mathematician then serving as the IDF's chief scientist, to suggest another way of photographing terrain across the canal: rigging a radio-controlled model aircraft (some were large enough) with a stabilized video camera. Dvoretzky took careful notes, remarking in passing that distorting thermals might be a problem. Three years later, in 1973, the state-owned Tadiran electronics house unveiled "Mastiff," the first of the small remotely piloted vehicles (RPV, now unmanned aerial vehicle or UAV, or simply drone), which was to launch what became an industry in which Israel remains a global leader. There is no evidence that the 1970 suggestion to Dvoretzky had any part in this (others were on to the same idea), but what happened was certainly indicative of the open-door mentality that so greatly facilitates innovation in the IDF.[9] That episode, moreover, exemplifies what creates the demand for innovation: the IDF respond to obstacles and setbacks by trying harder; and their doctrine strongly emphasizes the imperative of tenacity in pursuing the set objective—above all do not give up, especially when the situation seems hopeless.

Reservists with ideas have their opportunity when IDF leaders facing unpalatable alternatives look for new solutions for pressing problems, be they methods or techniques, hardware, software, or anything else. They will naturally turn to reservists they think might be well placed to provide answers.

Reservists who are managers in Israel's more efficient enterprises in their civilian lives are a natural source of managerial advice, and of course they themselves are forever battling to reduce the inevitable efficiency chasm between their own outfits and the IDF, whose built-in inefficiency is the inevitable consequence of abundant, almost free manpower mostly still in training, of equipment inventories forever sliding into obsolescence, and of facilities chronically underutilized except in war.

Other reservists who are not high-tech jet-setters or management gurus nevertheless enrich the IDF with more prosaic forms of expertise: for example, it sometimes happens when visiting an outpost that decent, even quite good food is served out of the very rudimentary field kitchen, cooked by a reservist on his annual recall who is a foodie or even a chef in private life. Other reservists on sentry duty might be the CEOs of important companies. Author Eitan Shamir at one point served in the headquarters of a reserve armored division alongside an older reservist on his annual recall, a sergeant whose job it was to gather and present maps, aerial photos, and other planning aids for the G3 operations officers—a humble clerical job. When the division's commanding officer found out that the sergeant was the CEO of a major Israeli company in his civilian life, he reassigned him from his humble duties to a much more important position to use his managerial skills. The CEO / sergeant was far from pleased—he had enjoyed his annual holiday from having to make hard decisions.

One can easily visualize what else IDF reservists transmit when doing their annual recall service, with so many young and inexperienced conscripts and so few career officers around them: all manner of expertise down to the most minute combat techniques. And of course reservists transmit to the conscripts their experienced, not to say cynical, sense of military values: look at us, here we are, still being recalled to serve years after leaving the standing army—you are needed now and you will be needed in the future. When your officer runs ahead calling out "*aharai*"—"follow me"—you will want to follow in attacking, or if it is a matter of standing and resisting, to do so. But do not look for opportunities to become a hero—it is not with muscle and blood that the IDF solve their problems but with *sechel,* literally "brains," but more accurately, discernment, a blend of intelligence and experience.

Sometimes only brute force will do, and certainly without a capacity for brute force there is no survival. But in the IDF—which are inevitably and uniquely civilian armed forces because of the centrality of their reservists—

it is not brute force, nor standing operating procedures, nor the traditions of the service, but *sechel* that comes up as the desired remedy in a tight spot. And that is how innovation also begins, out of an elemental impulse to look for entirely new solutions when old ones are ineffective or too costly.

This is the legacy of the IDF's founding generation. Moshe Dayan, Israel's one-eyed war leader (as both chief of staff from 1953 to 1958 and defense minister from 1967 to 1974), who had visited units of the US Marines in combat in 1966 in Vietnam, judged them to be "excellent fighters, bold, courageous, dauntless."[10] Nevertheless, Dayan recalled a day where the attendees of a diplomatic conference in Washington observed a parade by the ceremonial unit of the US Marines:

> I, too, clapped my hands in appreciation of their accomplished performance, but somewhere within me I felt a certain distaste, even anger and humiliation, at this use of combat troops as marionettes . . . The soldier's job is to fight, and does not do battle—at least not today—in straight and regular ranks and with fixed rhythmic movements . . . It is true that the fighting man is called a soldier and the men in an army wear uniform clothing, but battle demands of every man that he exert to the maximum his individual capability, and not that he moves his legs and swing his arms like a robot at the press of a button.[11]

Exerting "individual capability" to the maximum is the purpose of the IDF's most selective unit, Talpiot, nicknamed Sayeret Sechel ("brains commando") or, less kindly, Sayeret Chenonim ("nerd commando") (though leadership aptitude is a requirement).[12] It exists specifically to harness the creativity of young conscripts for the development of new military technologies. Its training course is the IDF's longest at forty-one months, five months more than the standard three years of compulsory conscript service, with graduates attaining both a bachelor of science degree from Hebrew University and the rank of first lieutenant, initiating a compulsory further six years as career officers, though some remain longer.

In the unit's recruitment process, selectivity is taken to an extreme: each year, the top 2 percent of all Israeli high school students are invited to take admission tests, followed by demanding exams in mathematics and physics. At that stage, some 200 candidates of the cohort are selected. Next, candidates undergo a battery of psychological and aptitude tests that further reduce their number to around 50 finalists, who can then start the program, with 40 or fewer successfully graduating.

Recruits are enrolled in a dual-degree course in mathematics and physics, which they are to study on a part-time basis (as do most Israeli students) while going through basic infantry training, followed by further training courses in every single branch of the IDF, thus variously serving as riflemen, tank gunners, gunner, radio operator, sailors, airmen, and more. In the process, trainees learn how the average soldier fares with the weapons and other gear issued in real-life conditions, as opposed to the controlled environments of laboratories and factories. While the emphasis is on technology (as well as the concurrent scientific education), the program does include courses in military doctrine and military history, to stimulate tactical and operational thinking in addition to technological problem-solving. While the overall aim is to stimulate independent thought and creativity, the participants are also given real-life assignments, such as the development of new specialized software that is actually needed by the IDF, the organization of a complex event, or the conduct of a university seminar in a specific subject.

Graduates of the program (Talpions), who number some 1,000, enjoy exceptional prestige in both the IDF and society at large. They include leading scientists and founders of some of Israel's most successful technology and biotech companies. The former director (2010–2016) of the Israel Defense Ministry's research and development department, BG Ofir Shoham, was a 1983 Talpiot graduate who commanded a missile boat at one point in his thirty years of IDF service, one of many and varied staff, command, and R&D assignments.

What the program as a whole has done for the IDF is to legitimize bottom-up creativity versus the top-down mentality generated by an inevitably hierarchical structure. Specifically, and of the greatest importance for innovation, the very existence of the Talpiot program affirms what is perhaps the most basic principle of the IDF as a whole: creativity outranks experience. It was for this reason that General Henry H. "Hap" Arnold, the talented World War II commander of the United States Army Air Forces, established Project RAND in March 1946, to provide new ideas for the unprecedented nuclear-weapon era that had just started and that, in his view, had invalidated the expertise accumulated by American airmen in the war just ended.

Talpiot's success led to the establishment of other elite programs, notably Shechakim ("Sky"), an intense program whose aim is to nurture cyberwizards and high-end coders. Another program, Havazalot ("Lilies"), prepares top candidates to become analysts in the intelligence-research division through a long and arduous journey that includes university-level studies of either Arabic or Persian as well as special in-house courses.[13] Program graduates are committed to serve for several years as salaried soldiers in addition to three years of conscript service.

The creativity that propels innovation naturally collides with deference to hierarchy. To effect change in the IDF, as in any organization, some existing arrangement, standing order, or established practice or priority must be changed, which can only happen if more senior officers are really willing to listen to subordinates with a new idea. The Talpiot program has had a significant role in the hierarchical reversal normally required for the accomplishment of any significant military innovation, but of course it is very small, so the insouciant attitude to rank diffused in the IDF by its many high-status but low-ranking reservists has a much larger role in facilitating innovation.

An extreme example, unimaginable in any other army, was witnessed in 1978 by author Edward Luttwak. He was visiting an especially exposed frontline outpost in southern Lebanon in the company of overall area commander MG Avigdor "Yanosh" Ben-Gal, then the IDF's top field commander and national hero following the epic victory of his Seventh Armored Brigade in October 1973. In that particular sector, the enemy's preferred tactic was to briefly bombard the outpost with mortar fire to force soldiers into their bombproof bunker and then immediately launch an infantry assault in the hope of overrunning the position before the soldiers could exit the bunker and run back to their firing positions along the perimeter. Because the enemy could also stop launching mortar bombs only to start again almost immediately in the hope of catching the soldiers still running out in the open between the bunker and their firing position, the standing order was to wear helmets and heavy flak jackets at all times.

After Ben-Gal and Luttwak arrived, because the real thing failed to happen there was a bunker-to-perimeter running drill. Ben-Gal noticed that one of the soldiers—a nineteen-year-old private—was not wearing his flak jacket. He called out to him to go back to the bunker and put it on immediately. The soldier instead stood before the general, who was especially reputed for his tactical expertise, to calmly explain that in his view it was more important to run across the open ground between the bunker and the protected firing position as quickly as possible, for which the flak jacket was a serious impediment, especially when exiting the shelter's narrow door. Ben-Gal replied that the standing order had been formulated by officers at headquarters after careful tactical study, and that he had better obey it, and right away. The soldier replied that he would obey, but only because he was outranked, still insisting as he went to the bunker to put on his flak jacket that the standing order was wrong, at least in the case of their specific outpost. A court martial, or at least a commander's summary punishment, might have ensued in some armies, but Ben-Gal merely joked that in the IDF even *pishers*—rude slang for "babes in diapers"—were sure they knew better than the general staff, while the twenty-one-year-old lieutenant in charge commented that the private was irritating and argumentative, but a pretty good soldier.

The episode was exceptional—even in the IDF privates do not normally argue tactics with senior field commanders—but there is a definite willingness to reinterpret orders, or even to ignore them altogether in obedience to a higher principle of officer responsibility officially accepted by all modern armies long before the IDF were established, but perhaps more often asserted in the IDF than elsewhere; that is, that officers must set aside orders to do the right thing, on their own responsibility.[14] As it happens, a significant IDF example again featured flak jackets. On June 10, 1982, Syrian commando units of the Eighty-Fifth and Sixty-Second brigades took up positions in Kafr Sill, a former hill village almost absorbed in greater Beirut.[15] Captain Doron Avital led his company of the 202nd Battalion, Thirty-Fifth Paratroopers Brigade, against the hilltop Syrian positions. Before the battle, an order came down from headquarters that all were to wear flak jackets. The paratroopers had been advancing until then against the ill-organized PLO (Palestine Liberation Organization), while the Syrians could lay down artillery fire to shield their positions with curtains of splinters, hence the order.

But Avital was convinced that the order would restrict his soldiers' mobility excessively, undermining their effectiveness. He repeatedly asked his commanders to reconsider the order but was refused. It was summer, the day was especially hot, and the company's battle plan called for a steep uphill march over the ridgeline to outflank the dug-in Syrian commandos. Avital told his soldiers that despite the order received he would fight without a flak jacket, but he left it to each soldier to decide whether to follow his example or not. Almost all did. In the ensuing battle, his company excelled, moving faster and maneuvering in more agile fashion than the other companies whose soldiers in flak jackets were exhausted by the long, steep climb.[16] Avital was not punished for disregarding the order; in fact he was promoted, later becoming commander of the IDF's most prestigious combat unit, the Sayeret Matkal.[17]

In the annals of the IDF there are many such stories because its culture does indeed sustain the principle of officer responsibility proclaimed by all armies, but which is usually rendered a dead letter by the failure to accept any disobedience, even if tactically justified. (Under Stalin, Red Army officers were shot for disregarding their orders no matter what, a reaction to the mass desertions and collapse of entire fronts in the summer of 1941.) It is evident that the prevailing IDF mentality that tolerates insubordinate applications of the principle of officer responsibility, and positively encourages the seizing of the initiative, must also favor innovation—even disruptive innovation that forces uncomfortable changes.

[5]

A Reserve Army of Innovators

IMPROVISATION, TACTICAL SKILLS, AND "FOLLOW-ME" COMBAT LEADership are not always enough, and then the lack of formal education and maturity can really show. That brings us to the central peculiarity of the IDF, which originated in a straightforward need for much more combat-ready military personnel than a standing army drawn from a small population could ever supply, but which incidentally provides the best possible remedy for the shortcomings of too-young would-be innovators.

When formally established by a decree of Prime Minister David Ben-Gurion on May 26, 1948, twelve days after the Declaration of Independence, the IDF already included the crucial innovation of a *reserve-centered* structure, whereby the training of personnel to staff reserve formations with stored weapons and equipment is a primary purpose of the active-duty ground forces (not many aircraft or warships could be provided for reserve air or naval forces), and not just a by-product of the training of the conscript intake.[1] Hence, the young conscripts who lack expertise and the too-few career officers are greatly outnumbered by the reservists who serve in all parts of the IDF, whose military expertise covers every branch, and whose civilian expertise ranges from astrophysics to zoology, as well as all forms of engineering, management, and more. And they, of course, keep feeding their ideas, good or bad, into the IDF.

Actually, the novel idea of a reserve-centered army emerged by a process of elimination.[2] When the time came in the years leading up to 1948 to deliberate about what kind of army the new state should have, the first to be ruled out was a professional army manned by full-time salaried soldiers and officers. That Roman invention, which remains the most common form of armies around the world, could not possibly work for Israel. With several Arab armies already invading its borders from the south, east, and north, it was obvious that the new state could not possibly field an army large enough from a total Jewish population of 650,000, as it was on May 15, 1948. Even if the funds could somehow be found, so many able-bodied men and women would be serving in uniform year in and year out that only the very young and the very old would remain to do any work.

An alternative was already in hand before the declaration of independence: a part-time volunteer militia, of which there were already two: the dominant Haganah and its rival the Irgun, the result of a 1931 political split. That was enough to persuade most people that the new state would need a *national* army, not politically motivated militias that would divide the nation instead of uniting it to confront its enemies—that being the Palestinian predicament even before the 1948 debacle, and ever since.

Part-time militias, manned by people coming and going as they please, could not be expected to absorb much combat training. Right up to the time when the IDF were formally established, most of the fighting was still simple infantry combat, or not even that actually—more a matter of odd groups of people occasionally shooting at each other with rifles, shotguns, and pistols. But as soon as the Arab states entered the war almost immediately after the declaration of independence on May 15, 1948, their handful of combat aircraft inflicted some damage and made a wholly disproportionate impression, while the few armored combat vehicles of the Egyptian Army, the Syrian Army, and the Arab Legion presented a terrible problem for the defenders, who had none of the myriad antiarmor weapons now present in every combat zone, but only a handful of PIATs (Projector Infantry Anti-Tank), a pathetic spring-propelled excuse for an antitank weapon that launched a hollow-charge warhead, inaccurately, over very short distances and that itself induced sheer despair.[3]

A United Press report of May 15, 1948, depicted the problem succinctly: "TEL AVIV. The battle for Palestine was on today with troops from Egypt,

Trans-Jordan, Syria, Lebanon, and Iraq converging against the Jewish defenders. Arab [Egyptian] planes bombed Tel Aviv three times. . . . Two Egyptian ground forces, with infantry and artillery, drove across the border . . . It's [they] spearheaded the Arab assault . . . crashing across the southern Palestine border at two points. Syrian and Lebanese troops roared down across the northern frontier . . . riding into battle in 150 armored trucks. King Abdullah of Trans-Jordan sent his Arab Legion and Iraqi regulars slicing across the eastern frontier . . . Jewish forces . . . captured Acre, most of Jerusalem and took over Haifa and Jaffa . . . Egyptian planes bombed the airport just north of [Tel Aviv] damaging one Air France plane . . . [The] Arab Legion . . . scored a major victory by capturing four Jewish settlements 10 miles south of Jerusalem . . . Jewish Haganah forces [launched] attacks of their own. . . . Irgun . . . forces . . . [captured] five Arab villages . . . Egyptian planes . . . showered leaflets . . . demanding that the Jews surrender. Reports on the Jewish defeat at Kfar Ezion . . . by [Transjordan's] Arab Legion . . . About 100 defenders were believed killed."[4]

In other words, as of May 15, 1948, the Jews could defeat the local Arabs with their superior training and cohesion even if all they had were assorted small arms. But they could not fight the invading Arab military forces with small arms alone; even the least formidable of the armored fighting vehicles they faced, the Arab Legion's 4×4 Marmon-Herrington Mk IVF armored cars, very much less than even a light tank, had enough armor to stop rifle bullets, while its two-pounder 40 mm cannon could demolish walls and improvised fortifications.

It became immediately obvious to all—and some had been saying so for years—that the IDF would need heavy weapons and competent crews to operate them, which could only be provided by prolonged and intensive training, much beyond the capacity of a part-time volunteer militia. With the two alternatives of a part-time militia or full-time regular army both ruled out and Israel lacking enough population for a conscript army of adequate size, the answer had to be an original invention: an army largely composed of fully equipped reserve units staffed, when necessary, by former conscripts recalled to active duty. Until then, those reservists could lead ordinary and productive civilian lives. That way, the country could field a disproportionately large army in wartime while otherwise keeping in uniform only the current crop of conscripts, as well as a relatively small cadre of professional officers and assorted specialists. The unavoidable cost was to provide complete equipment sets for each reserve formation—a great mass of expensive weapons, vehicles, communication gear, and more that would have to be maintained year-round, and periodically updated with new items of equipment.

There was nothing original about placing conscripts on the rolls of reserve units after their term of national service, a practice already well established in European armies in the nineteenth century, to add mass to their regular standing forces upon mobilization for war. Reserve forces long predated the outbreak of the First World War in 1914, when reserve regiments and divisions manned by former conscripts recalled for wartime service made up a large part of both the French and German armies. Also well established was the role of individual reservists, previously trained as conscripts or as short-service professionals, who could be recalled to active duty to serve in particular units.

Not even the *concept* of a reserve-centered army, that is one with more reserve than active-duty forces, was invented in Israel, because that had long been the Swiss model, one explicitly cited when the shape of a possible national army was being discussed in Israel's formative years, 1945–1948. Yet there was, and is, a critical difference. In Israel's case, war was already underway when the IDF were established in May 1948, so that the Swiss expectation that all reservists would be mobilized en masse when the nation was under attack—but not until then—did not apply at all. Instead, Israeli reservists would have to be called up to provide extra strength even between wars, to face the unending sequence of assorted security threats large and small in between major acts of war, with those serving in combat and support formations also recalled on an annual basis to keep up their combat-readiness with refresher training, and perhaps recalled again to learn to operate and maintain new equipment issued to their unit since its last annual recall.

What was entirely new in Israel was to establish an army *primarily* made up of reserve forces, staffed at all rank levels by recalled civilians. There were always too few first-line aircraft and naval vessels to allow the formation of large air or naval reserve forces, but when it came to the ground forces, the reserves greatly outnumbered active-duty formations. And during the 1967–1973 period of major wars, reserve formations were the main protagonists of the major battles, including the conquest of the Old City of Jerusalem in 1967 by the recalled reservists of the Fifty-Fifth Paratroops Brigade, the crossing of the Suez Canal in October 1973 by the 143rd and 162nd reserve armored divisions, and the counteroffensive on the Golan Heights by the reserve 146th and 210th armored divisions.

To ensure that they maintained the required skills, reservists had to be trained very thoroughly in the first place when they were still full-time conscripts, and then recalled for annual refresher training stints lasting as long as a month. That has always imposed a heavy sacrifice, especially on professionals and small-business owners forced to abandon their customers for weeks on end, but it has also kept the IDF on their functional toes—civilians taken away from their families and professional lives have very little patience with useless drills and poorly run exercises.

In any case, the IDF's high ratio of reservists to active-duty personnel connects army and society at every level, and in every way to a unique extent, especially now that universal military service has been abandoned in almost every other country.[5] IDF reservists, moreover, do not await their refresher training recall to reconnect with their unit if they have some innovation to suggest. Serving with them repeatedly over the years, reservists are on a first-name basis with their commanders and with their promoted former commanders serving in higher headquarters. That makes it easy for suggestions to reach the right address.

Unsurprisingly, reservists have been important innovators in the IDF. As early as 1954, mathematician reservists working on the Weizac (Weizmann Automatic Computer), Israel's first computer, and one of the first large-scale stored-program electronic computers anywhere in the world, made sure that some at least of the army's leaders understood the military potential of computers. By 1959 the IDF had established a computer unit with a US-made Philco.[6] It also started a course to teach suitable recruits programming skills, a modest one-classroom affair that has grown over the decades into a full-scale educational establishment that is now one of the world's leading sources of high-quality software professionals.

That first Philco machine, the TRANSAC S-2000, recently introduced with transistors instead of vacuum tubes (its supremacy lasted until the 1964 arrival of IBM's first 360), was first used to maintain up-to-date personnel records the reserve-centered IDF needed much more than most armies. Then it found more military applications, starting an IDF tradition of using computer capabilities in novel ways. Given that in 1959 hardly any IDF generals had any kind of higher education (some had not even completed secondary school), and that hardly anyone anywhere in the world knew anything about computers, it is obvious that the IDF might have remained computerless for years had it not been for its mathematician-reservists working at the Weizmann Institute.

The innovation connection can be very direct: IDF reservists, not only scientist-reservists or engineer-reservists, who are unhappy about whatever is issued by way of weapons or equipment of any kind, or the lack of some item of equipment that perhaps only exists in their minds, frequently initiate proposals within the IDF and beyond, perhaps contacting research centers of one sort or another, or one of the country's aerospace and defense firms. They can always find a fellow reservist on the inside, or at least the friend of a fellow reservist. Then they start pressing from the outside and keep at it until the relevant IDF command either takes up the evaluation or turns it down—and even then, they might keep trying. Insistence is in the Israeli DNA. Unless the reservists acquire a reputation as useless bores, doors are always open for them at the command headquarters of their own unit, and from there it is rarely hard to reach anyone throughout the IDF, top generals included.

[6]

A Different Military-Industrial Complex

WHEN IT COMES TO THE DEVELOPMENT AND PRODUCTION OF NEW weapons, new platforms, and new systems, the IDF benefit from a uniquely close relationship with the country's aerospace and military industries because of the dominance of its reservists in their management, research departments, and labor force.[1] Regardless of the ownership of the different firms, some entirely private, others entirely state owned, and some in-between, employees cannot ever forget that what they design, develop, and produce may be exported (as is increasingly the case) but will also be used by themselves if mobilized for war or by their sons and daughters—and that adds a compelling special meaning to the term cost-effective insofar as better equipment can reduce casualties. An engineering development project leader who served as commander of a reconnaissance detachment during the First Lebanon War in 1982 said that during the war he was sent behind enemy lines to observe and report, a very risky mission. "I thought to myself," he said, "there must be another way to be able to look 'on the other side of the hill' without exposing soldiers to such a risk."[2] When he came back from the war, he initiated the development of a series of tactical drones to accomplish the reconnaissance tasks without risking soldiers.

Aside from the emotional aspect, there is the sheer *quality* of the communications between the IDF customer and military suppliers staffed in great part by IDF reservists. All over the world such communications must be carefully calculated and guarded, even secretive, because huge amounts of money are involved, because any military acquisition can easily become politically controversial, and because, where corruption is absent, there are a great many bureaucratic rules to guard against improprieties and ensure strict impartiality—a factor especially important in the United States, where almost any major purchase can trigger lawsuits by the disappointed contenders.

Under whatever terminology, military purchasing anywhere begins with a request for proposals (RFP), which triggers intense efforts by would-be suppliers to understand what the military customer *really* wants. That is crucial because it happens not rarely that the military purchaser is not allowed to freely specify what it wants because the defense ministry would object for political reasons, or the finance or industry ministry object for industrial reasons, so that the military purchaser must compromise its own preferences to accommodate broader defense, financial, or industrial priorities. The major navies, for example, almost invariably prioritize "air-capable" ships that resemble real aircraft carriers as much as possible. Defense ministry officials on the other hand are apt to fear the size/cost growth of anything with "air" in its name, so they might force the navy to issue a request for proposals for a classic destroyer. But the offer of a warship that just happens to have a big deck with a hangar behind it is more likely to win a contract than one offering a true destroyer with weapons fore and aft and a modest helicopter pad. (Indeed, that is how the label "through-deck destroyer" was invented—to describe what was in a reality a small aircraft carrier).[3]

The above is a highly simplified version of a very difficult problem—the terminological traps are usually more subtle, making interpretation more difficult. Often, obscure details in an RFP are designed to favor a specific contractor—usually one's own traditional domestic supplier in the European Union countries. They are all supposed to select a European Union supplier strictly on merit, but the larger countries typically have their own entrenched local supplier, which is often a resonant national institution—Beretta for pistols and rifles, Rolls Royce for jet engines, Dassault for aviation, Kraus-Maffei for tanks and other armored vehicles—all faithful suppliers, and less sentimentally, providers of well-paid jobs for retired military officers and procurement officials. European efforts to achieve scale and efficiency in military procurement have been intense and have required countless conferences in such cities as Paris and Venice, and numberless working lunches and even harder-working dinners, but results have been elusive—the same national suppliers are still very much in business, supplying their own captive markets with equipment good, bad, or worse.

In the United States everything is different of course because, as everybody knows, the United States is ruled by laws. One of those laws happens to be the Buy America Act to discourage foreign competition in defense procurement, but that still leaves at least two contenders for any RFP, and their lobbyists, friendly home-state members of Congress, Washington offices, and consultants fight it out with a regiment of lawyers on each side ready to challenge any adverse decisions.[4]

In both the European Union and the United States therefore, there is no possibility of a frank, full, and continuous dialogue between military buyers and industrial suppliers. That is very unfortunate because the design of weapon systems rests on technologies that may change quickly, and the performance required of them may change quickly as well because of the arrival of something different on the other side, or because expected conflict patterns have changed. Hence, the entire weapon development process should be very fluid to accommodate changes quickly enough; otherwise, by the time the weapon system is actually produced and delivered to the armed forces, it may no longer fit changed military requirements, or it may even be obsolescent. But any such fluidity is drastically restricted by contractual obligations and detailed specifications that cannot just be changed as needed, without formal and very detailed "change orders" that require elaborate renegotiations, conducted by exchanging legally vetted correspondence, so that lawyers participate at each step, not just engineers and cost analysts.

That makes every agreed change very slow—slower still if there are disagreements, limiting greatly what can be done to keep weapon systems under development up to date. While the replacement of old-design components with better new ones may seem mere common sense, it still requires the renegotiation of the contract—without it, nothing can happen, and what does happen is indeed mostly nothing, because everybody is so fearful of reopening procurement contracts (perhaps triggering demands for a competition by the unsuccessful contender) that as the development process continues, more and more components slide into obsolescence as months give way to years, or even decades. That explains how it can happen that ultra-advanced jet fighters, handed over with great fanfare, can arrive for squadron service with some electronic components less capable than those found in some contemporary toys. Worse, the entire design may no longer satisfy military requirements, even though it had been ultramodern fifteen or twenty years earlier when the contract was signed.

In Israel, by contrast, the active-duty IDF people doing the buying and the mostly reserve IDF people doing the selling—or rather the research, development, fabrication, testing, evaluation, modification, and retesting—do actually talk to each other all the time, before and after the actual contract is signed, with no lawyers in between except when all the work has been done and final contracts can be drawn up. There is no waiting for contracts to be renegotiated to pursue design changes—which are essential to keep everything up to date as many components change—and there is no waiting for periodic progress reviews. There are instead informal channels of coordination between the IDF unit doing the buying and the industrial teams working to develop or actually produce equipment, with a single coordinating institution in the middle: the Administration for the Development of Weapons and Technological Infrastructure (known by its Hebrew acronym MaFat) jointly formed by the Ministry of Defense and the IDF to coordinate them and all state-owned entities engaged in the R&D and production of IDF equipment: Israel Military Industries (IMI), Israel Aerospace Industries (IAI), Rafael Advanced Defense Systems, the Institute for Biological Research, and the Space Agency.

The MaFat director, a full member of the General Staff, is a brigadier general, but his civilian suit indicates the hybrid position. MaFat's brief is to preserve the IDF's qualitative edge in weapons and infrastructures by directing domestic R&D projects and joint projects with foreign partners, and by nurturing exceptional manpower for them, not least through the Talpiot program. Its internal structure reflects the diversity of the disciplines MaFat is supposed to coordinate: applied science is the province of its Technological Infrastructure and Research Unit, which is supposed to supply useful applications for the R&D projects it has prioritized; the Space Administration is tasked with the R&D, manufacture, launch, orbit placement, and subsequent operation of all satellites; the missile-defense directorate overviews all antimissile R&D projects in cooperation with its US counterpart, the Department of Defense's Missile Defense Agency (MDA); the drone directorate's brief is to advance UAV capabilities and technologies; and a variety of other units are in charge of budgeting, one-off projects, and liaison with foreign partners.[5]

All this suggests a classic, multilayered, stovepiped, and compartmented bureaucracy, properly structured to do all the things bureaucrats like to do, which is to read papers and move them on from office to office, without actually doing anything very much that might go wrong and attract criticism. But so far, MaFat has been no such thing because at the time of writing its head (since 2016) is the notoriously headstrong antibureaucrat Brigadier General (Res.) Dr. Daniel Gold, a Talpion whose imperious direction of the Iron Dome project became the subject of a full-scale investigation, which condemned his countless administrative violations while recognizing he had achieved wonderful results, with miraculous rapidity, at very low cost.

MaFat was born out of a 1971 debate between the IDF and the Ministry of Defense over the fire-control system needed for what would eventually become the new Merkava battle tank—Israel's first such effort. Then the question was whether it would have to be imported, as with the diesel engine whose local design and production was unimaginable, or whether Israel's fledgling electronic industry was up to the task.[6] The unexpected outcome of this debate, which featured the Office of the Chief Scientist of the Ministry of Defense on one side and the IDF's weapon development department on the other, was a decision to merge the two into a joint civilian-military R&D unit (and also to go ahead with a domestic fire-control apparatus for the tank). This was the forerunner of what became MaFat in 1982 when Minister of Defense Ariel Sharon added the procurement and manufacturing directorates to the joint R&D unit. The great innovation in MaFat's structure lies in its hybrid nature—its head attends both IDF General Staff meetings as if he were a general, and the Defense Ministry's staff meetings of department chiefs as if he were an administrator. Indeed, he is both a general in the reserves and a civilian administrator.[7]

Another part of the defense establishment, Rafael, also experienced a revolutionary metamorphosis, in fact several. It was formed early in 1948 as the Science Corps to gather individual scientists to try to invent things that could help the desperately overtaxed fighting units in one way or other—but no miracles are recorded. In 1952, with a bit of funding, the Science Corps became the Research and Design Directorate, with both a research element and weapons-development unit. It was reorganized in 1958 as Rafael, after the Hebrew acronym for "Authority for the Development of Armaments," and later further renamed as the current Rafael Advanced Defense Systems Ltd., incorporated as a limited company in 2002. Even though still entirely state owned, as a self-standing company it can be a fair competitor for the country's private companies. But by then Rafael's transformation into a veritable innovation machine had already been accomplished: still a very small organization as compared to its peers, it had somehow developed a slew of new weapons: the Python series of air-to-air missiles; the Spike family of fire-and-forget surface-to-surface missiles; the Popeye very long-range air-to-ground missile, which is believed to be the basis of a nuclear-armed, submarine-launched cruise missile; the Iron Dome system for the low-cost interception of even cheaper rockets, but expensive missiles also; Trophy, the first active defense system for armored vehicles in the West (preceded by Russia's Drozd and Arena), the world's first operational unmanned surface vehicle; and David's Sling, an antimissile system of much longer range than Iron Dome.

If any one individual can be credited for transforming a group of worthy scientists and devoted administrators who had already accomplished much (including the development of the country's first and remarkable air-to-air missile, Shafrir), it was Moshe "Musa" Peled, a retired major general and an odd exception to the engineers who preceded and followed him in the job. An officer who had played a large role in turning heroic defense into sweeping victory on the Golan Heights in October 1973 as commander of an armored division of reservists equipped with mostly upgraded World War Two–vintage Sherman tanks, the long-retired Peled seemed a strange choice when he was appointed president of Rafael in 1987. It was only in the world of armor officers, including Americans, Europeans, and Russians, that his 1973 campaign was widely recognized as a true classic—an offensive that drove back across the border much larger Syrian forces equipped with more and better tanks, by sheer dynamic momentum, kept up by responding to any pause caused by exhaustion or enemy resistance by sending forward any force in hand, large or small, to keep up the drive.[8]

In his new post, Peled mounted another persistent offensive, this time against bureaucratic proclivities and tendencies to play it safe. He did not think it worthwhile to pursue incremental innovation—of the kind that accounts for 90 percent or more of R&D spending worldwide—which minimizes the risks of failure by sticking to the improvement of existing platforms and weapons, but surrenders any chance of real breakthroughs. Peled demanded macroinnovation or nothing, high-risk leaps into the truly new, which might fail of course. He reportedly told his engineers: "If every project results in a success it means you are not daring enough. I would expect an overall failure rate of 50%."[9]

That was Peled's way of fighting any relapse into mediocre normality at a time when he feared that Rafael might lose its poverty-sharpened edge because its exports were bringing in much more money than ever before, adding to rising IDF funding. It had acquired its first decent headquarters building complete with new furniture, a far cry from a past of extreme scarcity. Peled's offensive was successful: instead of sliding into well-funded corporate mediocrity, Rafael became more of a risk taker than ever before, and not only technologically: its managers went out on the longest possible limb to engineer-develop the Iron Dome system before any government funding had been authorized.

[7]

High-Speed Development

From Missile Boats to Iron Dome

TWO MISSILE SYSTEMS DEVELOPED DECADES APART ILLUSTRATE HOW Israel's acquisition process differs from the normal way of doing such things, chiefly by a much faster pace, as risks are accepted in preference to the high costs and endless delays of adversarial testing and verification procedures. The first was the antiship missile Gabriel, developed, along with its fire-control radar and guidance system, by the country's infant military industries from 1962, when Israel had less than half the population of Sicily—and very little by way of electrical or mechanical industry, hardly more than a few small machine-tool shops, repair garages, smithies, and such. The second, Israel's largest weapon-development failure, was the Lavi jet fighter, condemned not by rapid development but by US opposition; begun in 1980, it was canceled in August 1987 because of US pressure after two prototypes had flown.[1]

Aside from poverty, and the country was indeed poor by European, let alone American standards, the other cause of industrial underdevelopment was Zionist ideology: it greatly favored agriculture ("to redeem the neglected land with manual labor," which would in turn redeem the weakling Jews of the diaspora). A majority of Israel's political leaders, academics, and opinionmakers despised commerce; from banking to shopkeeping, it was all equally sordid for them, and they disregarded industry because their great dream was to advance Israel's agriculture. In that they were quickly successful—Israel became a leader in the development of new crops, in the use of bromine for fumigation, and in globally successful drip irrigation. In the meantime, industry lagged, starved of enthusiasm and leadership.

It was in that most unpromising context that the dream of a missile-boat (only the Soviet Navy had any) had its very uncertain start. At the time, Israel's Research and Design Directorate amounted to a few engineers in a few ex-British Army huts, with little funding or equipment, but it soon embarked on its first attempt to develop a tactical missile, the surface-to-surface Luz, which was to be manually command-guided with a joystick and offered to the artillery and the air force.[2] Development was stopped in 1963 as both branches considered it unsuitable for their needs, but in 1964, the IDF's Sea Corps adopted the project together with the Israel Aviation Company, now Israel Aerospace Industries, when it realized the Egyptian and Syrian navies were acquiring Soviet Osa-class missile boats equipped with the powerful P15 Styx antiship missile.

The start was unpromising because the semiactive radar guidance of what became the Gabriel missile required the uninterrupted radar illumination of a target by the launching vessel's radar, hard to do with both launching vessel and target vessel on the move. At first it could only be done at short ranges, ten kilometers or less. But the Gabriel's leading artificer, engineer Uri Even-Tov, was a master of improvisations and go-arounds.[3]

To build a useful capability against the Osa missile boats, two development efforts were launched, each of might be called major were it not for the handful of engineers and scant means at hand. The first effort was to improve the missile motor to attain a more useful range of twenty kilometers, a task assigned to its developer, Israel Military Industries (IMI), which amounted to a few workshops with a few engineers. The second effort was to develop an active radar seeker that could keep up with the target even if it maneuvered evasively.

The resulting missile design envisaged three stages: first, for the initial takeoff boost, it was directed manually with left and right commands; second, the missile's own semiactive radar took over, with its returns processed aboard the launching ship to generate guidance commands; third, a few kilometers from the target, the missile dove down into a sea-skimming trajectory and switched on an active seeker to guide it to the target. That front-mounted radar receiver would continually detect signals bouncing off the target vessel once it was illuminated (aka "painted") by the onboard radar transmitter and feed the inputs to an autopilot that would in turn generate directional signals for the moving fins that sent the missile right or left as needed. Its up-or-down movement was regulated by an altimeter to achieve the sea-skimming flight profile in order to minimize the missile's exposure to visual or radar detection. The Israeli Navy had not wanted a large missile like the Styx that flew like a dive-bombing airplane and could be shot down like one, so right from the start it insisted on a much smaller sea-skimmer that would be hard to detect in all the sea clutter that shows up on maritime radar scopes.

Initial development of the Gabriel was very slow because of lack of funds. The ground-force generals who dominated the General Staff agreed with the airmen, who were themselves convinced they could sink any ship with their fighter-bombers without need of naval missiles. But on October 21, 1967, with a cease-fire in place after the June 1967 war that left Israel in control of the Sinai, the 1944 British-built destroyer escort ex-*MS Zealous,* serving as the flagship of the Israeli Navy as the *INS Eilat,* was sunk in international waters off Egypt's Port Said, hit by three Soviet-made Styx missiles launched from Komar-class missile boats of the Egyptian Navy from inside the harbor. Out of a crew of 199, a total of 47 were killed and more than 90 wounded. Sixty-seven hours after the attack Israel retaliated by shelling Port Suez with mortars, destroying two oil refineries. That episode was enough to change the priorities of the IDF General Staff. With sudden funding, and a furious day- and night effort, the Gabriel's development was largely completed by the end of 1969. Four years later during the October 1973 war it was devastatingly effective, rapidly sinking seven Egyptian and Syrian warships and driving others to seek refuge within their ports.

The high-speed development of the Gabriel system—the missile, radar, fire-control system, and an entire suite of electronic countermeasures that largely neutralized the guidance of the Soviet Styx missiles of the two Arab navies—was an extraordinary engineering achievement. It reflected a specific operating concept developed beforehand by the chief guiding light of the Israeli Navy, Rear Admiral Yohai Ben-Nun, who had become at age thirty-six commander of the Sea Corps, Israel's navy in 1960.[4] Yohai Ben-Nun was unimpressed by contemporary warships, destroyers, and cruisers that needed large tonnages to carry guns of limited range, with large crews because of operating requirements and more requirements generated by crew needs. Those classic warships were fine to show the flag, but long before the *Eilat* was sunk, Ben-Nun concluded that they were vulnerable to drastically smaller missile boats. He wanted a navy of small, fast boats to sink much larger traditional warships.

One of Israel's first combat frogmen, personally trained in secret under British rule by a veteran of Italy's world-leading underwater combat force, Ben-Nun founded the Shayetet ("Flotilla") 13 naval commando force in 1949. He secretly imported leftover wartime Italian equipment, including manned torpedoes, with the connivance of the Italian frogman unit that survived on its traditional base in semiclandestine fashion because of postwar political restrictions.[5] Unable to match the larger Arab navies at sea—Israel could not afford proper modern warships—it would instead send Shayetet's combat frogmen to attack and sink their warships at their anchorages inside their guarded naval bases with limpet mines and manned torpedoes, as the Italians had famously done against the Royal Navy in Alexandria and Gibraltar.[6]

By the time the IDF's Sea Corps came under Ben-Nun's command in 1960, he had come up with an entirely different way of overcoming its fundamental problem, namely the impossibility of acquiring and operating even a handful of modern warships, given that the air force had to come first in funding, followed by the mechanized forces essential to protect Israel from invasion. Aside from minesweepers, torpedo boats, and other specialized vessels, there was a hierarchy among surface warships, then still very much a function of the power of their guns. It started with corvettes of a thousand tons or so of displacement, rising to 2,000-ton-or-so destroyer-escorts, 3,000-ton destroyers, and 5,000–6,000-ton cruisers, each class capable of mounting bigger and bigger guns. Above them were even more impossibly large warships, all the way up to aircraft carriers, as well as submarines that might have small displacements but were drastically more expensive per ton.

Given that he could not hope to acquire modern corvettes, Ben-Nun's solution was to give up on conventional warships and their guns altogether to instead build a navy of antiship missile boats that could achieve ship-sinking lethality within displacements under 500 tons, even under 300 tons. Unlike the long-established torpedo boats, which were strictly short-range attack vessels, missile boats could have decent range and endurance because their major weapon was not heavy.[7] But Ben-Nun's problem was that no such missiles or vessels existed. For the missile, the Gabriel program begun in late 1962 would offer the solution, incidentally launching Israel's missile industry, which started in Rafael's few huts and evolved into today's supplier of advanced air-to-air, air-to-ground, and surface-to-surface ballistic and antiballistic missiles.

But for Gabriel's platform a do-it-yourself vessel was simply impossible. At the time, Israel's only shipyard could at best repair vessels and build barges. Just as he had done for his frogmen when he had reached out to Italy's pioneers, Ben-Nun sent a trusted representative to scout shipyards throughout Europe. What he wanted was a vessel that could compress the needed range, speed, and load capacity within a small tonnage, on the order of 200 tons at full displacement, which would allow for the purchase of a dozen such boats for the price of one small destroyer.

A newly revived West German shipbuilder produced the best hull design, and the best engines were also German,[8] based on the German-designed and German-powered Jaguar torpedo boat, but with an Israeli-specified steel hull instead of the original wooden hull, the addition of 2.4 meters (seven feet, ten inches) to the hull length, and revised internal compartments. The Gabriel missile launchers and other weapons and related systems of what was designated as the Sa'ar 3 class were to be installed in Israel upon the arrival of the boats.[9]

There was historical irony in the fact that the Jaguar, derived from the famously fast S-boot of the wartime German Navy, was the most suited to function as a platform for the Israeli Navy. Only fifteen years after the end of the war and the Holocaust, Israeli Navy officers found themselves engaged in vigorous technical discussions with German officers and engineers who had served in Nazi uniforms, even encountering heavily compromised figures.[10]

Once a development agreement was signed, an Israeli delegation arrived at the German shipyard to discuss needed modifications under conditions of extreme secrecy because diplomatic relations did not start with West Germany until 1965.[11] The modifications were extensive; in addition to replacing the wood hull with steel, a longer frame was needed to accommodate the weapons and electronic systems, as was a different mast. When the head of the Israeli delegation finished presenting the long list of Israeli requirements to the Germans, there was a moment of silence, after which the German team chief asked, "Sir, please, would you like to also have a grand piano on deck?"[12]

The Germans balked at accommodating all the changes the Israelis wanted but when everything seemed blocked, Israel's delegation head, Rear Admiral Shlomo Erell, asked to deal directly with the chief engineer. When the reluctant Germans finally agreed, the chemistry between the two men quickly advanced the design to meet Israeli specifications.[13] The shipyard agreed to supply Israel with twelve Jaguar-class fast attack craft, but only three had been delivered when in 1964 the West German government repudiated the agreement under Arab diplomatic and commercial pressure, though it did agree that its fully engineered design could be built in a foreign yard. With France the only country ready to supply Israel, the choice for the remaining nine boats fell on Constructions Mécaniques de Normandie, a small private shipyard in Cherbourg, under the class designation La Combattante (designated Sa'ar 1 by the Israelis). The first boat for Israel left in April 1967 and another a month later. But on June 2, 1967, just a few days before war started on June 5, President Charles de Gaulle declared that France would no longer supply "offensive" weapons to the Middle East, which meant Israel, there being no Arab buyers of French weapons at the time. All nine boats had been paid for and the remaining seven were in advanced construction, with two almost ready. They duly sailed for Israel in the autumn of 1967, unresisted; many in the French defense establishment had been comrades in arms with the Israelis since the 1950s, and Cherbourg's population was also strongly supportive.

De Gaulle's embargo became total after a clamorous Israeli raid on the Beirut airport on December 28, 1967, but Parisian policy decisions were again subverted locally, and three almost-completed missile boats embarked with reduced crews on January 4, 1969. As they sailed out to practice, as before, they raised the Israeli flag and continued unchallenged into the English Channel, never to return. It caused a scandal in Paris, which ordered increased vigilance, but construction was allowed to continue on the last five missile boats. However, Israeli crew members allowed aboard to practice were kept under surveillance, and there was a tight control on fuel loadings, with the French Navy alerted to stop any escape. De Gaulle resigned on April 28, 1969, but French policy did not change, and it took a full-scale Mossad operation with a Norwegian-purchase cover story to pull off the escape of the five remaining boats over Christmas, December 24–25, 1969. It was not a low-risk operation; that night French vigilance was relaxed due to a dangerous gale in the Bay of Biscay.

Because a BBC helicopter crew later filmed the escaping boats in the Channel, causing much hilarity around the world at this blatant defiance of the imperious de Gaulle, the French minister of defense and de Gaulle supporter Michel Debré ordered the French Air Force to find and sink the boats, then still at the start of their 3,145-nautical-mile (5,825-kilometer) journey to Gibraltar, Crete, and finally Haifa, with refueling along the way from improvised tankers. The French military chiefs, for whom the Israelis had been comrades in arms since 1956, were unwilling to add violence to betrayal and delayed their response to Debré's order until it was countermanded by Prime Minister Jacques Chaban-Delmas. He could have been overruled in turn by the just-installed president, Georges Pompidou, but Pompidou was not as devoted as Debré to de Gaulle, and in any case he was unwilling to challenge Chaban-Delmas, a tennis and rugby champion and an authentic hero of the wartime resistance. It remained for French Foreign Minister Maurice Schumann to warn that if the boats appeared in Israel, "the consequences will be very grave indeed."[14] But the arrival of the boats in Haifa harbor on New Year's Eve, December 31, 1969, triggered a memorable citywide party.

Ben-Nun's foresight was vindicated less than four years later in the October 1973 war, when his missile boats and their Gabriel missiles dominated the war at sea.[15] In history's first battle between missile boats, on the night of October 6–7, 1973, the Israelis sank one Soviet-made Osa and two Komar missile boats, one torpedo boat, and one minelayer of the Syrian Navy near the naval base in Latakia's harbor. On the next night, they sank three Egyptian *Osa*-class missile boats in Dumayit harbor and badly damaged another. It was a baptism of fire for the new boats, plus the Israeli-built Sa'ar 4, locally made with a larger displacement and upgraded radar and systems.[16]

Iron Dome

When it comes to rapid development, the ultimate example is Iron Dome (Kippat Barzel), the antirocket, antimissile, anti–artillery shell, and potentially antiaircraft system whose full-scale development started in 2007, based on a research program begun in 2005. Declared operational in 2011, it achieved global fame by intercepting 421 rockets in the seven-day conflict in November 2012 with Hamas and another 578 rockets during the fifty-day confrontation in summer 2014.[17] Like any weapon system, Iron Dome has its limitations, but compared to the fifteen to twenty or more years that guided-missile projects take elsewhere, the sheer rapidity of its development—less than five years—was phenomenal, given that the radar detector and tracker, the remarkable software, and the interceptor missile were all entirely new.

The software was the key breakthrough because it made the system economical, allowing Israel to intercept endless rockets, most crudely and cheaply made (approximately US$500 each) yet still potentially destructive and possibly devastating, with missiles that could have cost at least a hundred times as much.[18]

Studying the history of rocket attacks, Israeli planners concluded that about 75 percent of the rockets would fall in open terrain and therefore did not warrant interception. Therefore, the Iron Dome's computer was to plot the trajectory of incoming rockets to assess their final hit-point and launch interceptor missiles only at those computed to fall on residential areas or on important civilian or military infrastructure. Operators needed the ability to override the software because even intercepted rockets can inflict damage; in spite of the mere seconds available, manually controlled intercepts are accomplished routinely, saving many lives as well as preventing much material damage.[19]

The Iron Dome is the ultimate evidence that Israel's ultrafast, often risk-taking acquisition process offers two very different benefits. One is obvious: the timely delivery of effective weapons urgently needed for recurring episodes of combat. The other is the economy of straightforward programs, amounting to vast savings. And cost is decisive for IDF innovations because in most cases if a new weapon or system cannot be developed on the cheap, it cannot be developed at all.

Haste also makes some waste, of course. To rush ahead with purchasing and fabrication before all the calculations are fully complete results in both false starts and backtracking. But even somewhat wasteful haste is economical compared to the cumulative cost of carefully planned, managed, costed, and executed development programs constantly overtaken by the obsolescence of components, changes in the nature of the threat, or technological change. The agonizingly detailed procedures obligatory in both the United States and Western Europe, which delay the acquisition of weapon systems for years and even decades, inflict tremendous costs while attempting to limit them by preventing overbilling, waste, technical fraud (fake test results), and mismanagement. Elaborate rules to ensure arm's-length contracting practices designed to prevent cozy deals between complaisant acquisition officers and greedy contractors require that everything must be costed out and specified in excruciating detail. That in itself requires a great many working hours by managers, accountants, and lawyers who routinely outnumber the engineers involved. Then there is more of the same for each successive design review, program evaluation, and cost analysis, which also require endless recalculations, often outsourced at great cost, increasing the total expenditure before anything is produced.

But the worst part is that during the years and decades of the development process, more and more components and even subsystems (in combat aircraft they can be subsystems as important as radar or even engines) become obsolete, or may even be going out of production in the near future. Each time that happens, the result is an agonizing dilemma: Change to a new item, which will usually require redesign and reengineering, typically triggering yet another time-consuming program review. Or else remain with the old, and thus contribute to the problem of obsolescence-on-delivery, which may finally trigger outright cancellation—that being the fate of many a costly weapon system—after huge sums were spent for naught.

Consider for example the costliest of all contemporary weapon acquisitions, the F-35 jet fighter family, whose development started in 1996. It was declared operational ("initial operational capability" or IOC) some twenty years later, in 2015 for the US Marine version, 2016 for the Air Force version, and 2019 for the Navy version—except that these were political IOCs to counter criticisms of the slow pace of development, and many inadequacies remained to be corrected.[20] Together with the US Air Force, the Israeli Air Force (IAF) has been spearheading the F-35's operational use, and the IAF was the first air force to use the plane in combat.[21] Even then, it flew with unresolved software problems. When an aviation project is so greatly protracted, dragging on for more than two decades, it is bound to be left behind by the rapid advance of microprocessor technology. Any combat aircraft is also a set of computers: while some microprocessors are easily updated, others are embedded in components and subsystems that have to be redesigned, delaying the project once again, while there are bound to be further advances. The F-35 sacrifices speed, maneuverability, and its air-to-ground bombloads in order to maximize its stealth characteristics, which would have been extremely valuable in combat in 1996, but are much less so now that both lower-frequency and bistatic radars can detect stealth aircraft in varied conditions.[22]

When it came to the development of the Iron Dome air-defense system from contract inception in February 2006 to its March 2011 IOC date, Israel was no longer the poor little country that had embarked on the development of its first missiles in the early 1960s. Yet there was extreme scarcity nonetheless, both in time, in the wake of that year's massive Hezbollah rocket bombardment (the total number of rockets was variously estimated at between 3,970 and 4,228 between July 12 and August 14, 2006), and even funds, albeit because of policy priorities rather than national limitations. With a greatly diminished threat of conventional war as compared to previous decades—owing to the withdrawal of Egypt and Jordan from the conflict, Syria's decline, and the incapacitation of Iraq—the funding needs of the ground forces and even the air force to a degree were less pressing. Nevertheless, their bureaucratic power was undiminished, and while all might acknowledge the need to defeat the rocket threat, no branch of the IDF was willing to cut its own budget to do so.

In any case almost all IDF senior officers were convinced, along with almost all civilian experts, that it would be disastrously uneconomical to intercept rockets that were virtually free for their users with necessarily costly interceptor missiles.[23] Others argued that the successful protection of its own population would delegitimize Israeli air attacks meant to deter or stop aerial bombardment because the casualties would all be in Gaza, which would result in international pressure against Israeli counteroffensive operations, as duly happened in the summer of 2014 and in May 2021.[24] The result was that Iron Dome's development process was at first only minimally funded; indeed it was funded illegally. A timeline best shows how that extraordinary situation came about.

On April 19, 2004, the IDF deputy chief of staff, MG Gabi Ashkenazi, assigned responsibility for the overall effort against the medium and long-range surface-to-surface missile threat to the air force. On July 6, 2004, he assigned all staff work to MaFat, the "Administration for the Development of Weapons and Technological Infrastructure." On October 13, 2004, Deputy Chief of Staff MG Dan Halutz (former commander of the Israeli Air Force) formally directed MaFat to explore possible countermeasures both against the short-range Qassam rockets launched from the Gaza Strip, and against longer-range rockets Iran was supplying to Hezbollah in Lebanon.

But Danny Gold, head of MaFat's research and development unit, was vehemently dissatisfied with mere exploration.[25] On or around August 5, 2005, he decided on his own to initiate accelerated development while the airmen were still undecided on how to proceed and the ground-force generals were opposed. In their view, rocket bombardments were not to be tolerated, nor mitigated with expensive attempts at interception—they were to be extinguished at the source by overrunning those who launched rockets. Gold was unimpressed by the argument: rocket attacks were continuing while nobody was doing any overrunning because the potential cost in blood, treasure, and political capital of ground incursions into Gaza outweighed the damage and casualties inflicted by the rockets. Meanwhile, Israeli towns and villages next to the border, such as Sderot, had to live under intermittent bombardment that made life excruciating and inflicted some casualties—90 percent of the residents experienced a rocket explosion on their own or an adjacent street.[26]

Gold did have the authority to pursue any research and development effort he deemed worthy, but only up to the prototype stage, which could be done within his small budget, because all sorts of liberties are allowed in making prototypes without having to worry about future producibility, continuing maintenance, or even reliability.[27] But from the start, he directed his subordinates to pursue the project as an all-out development effort whose costs would soon greatly exceed his budget, with the aim of demonstrating an intercept capability within eighteen months and completing the full-scale engineering development of each component—radar, software, and missile—within thirty-six months. Moreover, he wanted the equally rapid establishment of industrial production facilities, including assembly lines for a missile that did not yet exist and whose full-scale engineering development had not been authorized.

In the absence of miracles, this circle could only be squared with the next-best thing, the engineering equivalent of a commando raid: a "telescopic" project in which everything was done concurrently instead of sequentially, saving much time. But there was a price to be paid, because of the near certainty that the development of the radar, missile, and software would keep diverging as work evolved to overcome specific problems, requiring costly fixes that might also upend the desired urgency by imposing their own delays. On top of that was Gold's entirely unauthorized foray into the preparation of industrial facilities, for which he had no funds at all. Even his engineering development work was unauthorized, because it went far beyond mere prototyping with cheap mockups, brass-boarded components, and such, but the amounts were much smaller.

Invested with his own overwhelming sense of urgency—a determination to have defenses up and running before the next mass rocket bombardment—the charismatic Gold conjured the funds needed out of thin air by persuading the managers and board members of Rafael, the state-owned missile company, and Elta, the state-owned radar company (almost all of them reserve officers), to advance the money from their own funds. That they did, without having government orders for the work, and therefore without any certainty of repayment soon or ever, even though they were the directors of incorporated companies legally responsible for their financial integrity. They were risking legal retribution, their reputations, and their careers by going along with Gold, in order to continue straight into much more costly full-scale development after he had run out of properly authorized prototyping funds.

These highly irregular—indeed prohibited—procedures seem to have favored rather than hindered the project because all concerned were caught up in its overwhelming dynamic momentum. Gold's own eight-member MaFat team, Rafael's best missile designers, and Elta's top radar engineers all worked as near a 24/7 schedule as human physiology would allow. Everyone gave up private life for the duration, and at least one religious staff member gave up his holy Sabbath, invoking the allowed "saving of lives" exception.[28]

No less important was the question of method. Truly innovative research and development, as opposed to incremental upgrading that uses up most research and development funds, must be a process of trial and error, or rather of rigging up something, testing it properly, and then deciding how to correct or work around the shortcomings or outright defects that emerge. Hence some costs and some delays are unavoidable each time there is enough progress to warrant testing. But when there are commercial contractors at one end and government servants at the other, civil or military, the contractors' proper concern with profits and the government's duty to the taxpayer means that each error must first be analyzed carefully, in order to determine whether the error was caused by overdemanding specifications or excessively demanding testing, the government's fault, or by the contractor's inadequate skill, attention, or investment. It takes time, sometimes considerable time, to verify claims on each side, and it may generate disputes that bring lawyers into the process, so that dispute resolution takes up yet more time before work can resume to correct the problem or to come up with an alternative. Moreover, attempts to limit waste, fraud, and mismanagement through elaborately supervised verification procedures by an outside testing group, in order to prevent cozy arrangements at the taxpayers' expense, are a very expensive way of avoiding costs in new weapons development.

Gold and his team acted differently, or rather they acted just as US weapons developers used to do before many thousands of legally mandated regulations were imposed on them from the 1960s. When testing revealed an error, everyone started work right away to overcome the problem without bothering to attribute responsibility, let alone blame. Gold also mandated a "design to cost" method: the 400 or so staff from Elta, Rafael, and his own organization were divided into fourteen teams that functioned as parallel competitive start-ups: "We came up with various solutions in parallel—everything in the project was done in parallel."[29]

Any normal project comprises two stages: an R&D phase and an engineering development and production phase. Gold compressed the two stages to keep up with the tight deadline. This too was a decision that exceeded his authority, but he took the chance nevertheless. Actually, the key innovation of the Iron Dome system was the low unit cost of its missiles, between US$77,000 and US$97,000 initially and roughly US$50,000 as of 2021.[30] It means that this missile falls into the ammunition category, allowing a non-conservative firing doctrine that permits the IDF to take risks with lower-probability missile launches, thereby achieving a higher interception rate.

Paradoxically, the initial reluctance to acquire the system within the IDF greatly aided the project's development.[31] As a senior participant explained: "Because no operational demands or technical specifications were imposed on them, the engineers enjoyed total freedom to develop an optimal weapon rapidly . . . without [junior officers] pouring into it their 'wet dreams' derived from *Aviation Week & Space Technology* magazine, and without generals 'overseeing' the program with endless meetings and discussions."[32] Unintentionally, the disregard for proper bureaucratic procedures, later condemned by the state comptroller, actually saved the day.[33]

In the aftermath, it could be argued that a better performance might have been attained with more R&D prior to production, but the price would have been much higher. Gold's team was ingenious in identifying features where funds could be cut without damaging the final product.[34] Another key to cost control in the engineering phases was that Rafael's managers agreed to a predetermined fixed payment that would not be dependent on the amount of development time.[35] It meant that the company's managers accepted the full risk of losing money on the project. But it also meant they had full freedom of action because the customer could not intervene to reduce costs, which played a major role in the success of the project. In one case, they diverged radically from established practice by deciding that Iron Dome's missile should have an electrical servo motor, as if it were a large liquid-fuel ballistic missile. This decision ignored the expertise accumulated over the years that favored pneumatic servo motors for air-to-air missiles, but the unusual electrical solution considerably reduced overall costs.[36]

Pursuing low-cost solutions for each part of the system—and the Iron Dome is a veritable orchestra of subsystems—was a high priority because at the outset all the options seemed to lead to astronomical expense. Finding expensive, complex solutions in designing weapons is not hard—it is the realm of US$6,000 aircraft toilet seats—but finding simple solutions requires thought and creativity.[37] In spite of the urgency of finding a solution for the ever-present threat of rocket and missile bombardment, there was no compromise in meeting the fundamental requirement of operating efficacy at a reasonable price. When the developers of the separate components came up with solutions that failed to attain the required standards, they were told to keep working until they found better solutions. "The success of a project always depends on prior success in choosing the right people" was the conclusion of one protagonist. Dr. Ron, one of the development team leaders called the "Fabulous Five," describes the special atmosphere in the project: "From the first moment, I was driven by the fact that we were dealing with something of a supreme significance . . . A senior engineer with decades of experience found himself listening to a young engineer with a different opinion." Indeed, the absence of a hierarchy naturally inclined to favor well-established methods propelled the project. There was no higher authority to override the preference for the quickest, cheapest, and most logical operational system merely because it relied on unusual solutions.[38]

As happened repeatedly and on all sides during the Second World War, groups of engineers and scientists personally committed to an urgent national mission that might avert the deaths of loved ones achieved a critical mass of dynamic creativity otherwise not only unattainable but unimaginable. That is how the British came up with the world's first centralized air defense system to defeat the initially much superior Luftwaffe with nary a computer in sight, and then forced a crude computer into existence to break German radio traffic ciphers to sink the U-boats starving Britain. That is also how the Germans invented and produced the first rocket-powered fighter, the first jet fighter, the first cruise missile, the first ballistic missile, the first air-to-surface missile, and even the prototype of the first stealthy jet bomber while under increasingly heavy bombing. All those achievements were exceeded by the mostly refugee scientists of the Manhattan Project, who feared that German creativity might extend to the fission bomb and therefore stopped at nothing to invent practical ways of separating uranium-238 and using the resulting fissile material in two different bomb designs, hoping one might work (both did), all done in a little over three years.

In Israel's case we have the testimony of the head of Rafael's R&D Division, Dr. Ronen: "I was traveling north one day to attend an experiment and when I reached Tel Aviv an alarm was sounded. My children and grandchildren live there, and I witnessed how the missile intercepted a rocket seemingly right over their house." David, the project's chief engineer, explained how the project dynamics worked. Before Iron Dome, the rule within Rafael was that after a failed test, everyone would go back to the development facility to return to the proving grounds with a new solution within a few months; the Iron Dome project broke that rule. First, the experiments started very early, which is highly irregular in missile development. Right from the first, frustrating experiments, an ethos of "no surrender" was forged. Sometimes a team would remain in the field instead of returning to their desks while the guys at the labs figured out the problems; sometimes the solutions were found on the same night that an obstacle arose, so that the following morning a new experiment could take place as if no mishap had ever occurred. In the process, all barriers to creativity were breached, including the most important, those that were entirely unconscious. A key figure testified: "The [Iron Dome] . . . missile is the only one in the world that contains components [taken from toys]. One day I brought to work one of my son's toy cars. We passed it between us and saw that there were components that really suited us; more than that I cannot tell you."[39]

It is an unfortunate hallmark of military R&D to reinvent what already exists, often to find a use for some new technology that may excite technologists but is not really necessary, or simply too expensive, for the project at hand. The aircraft carrier *USS Gerald R. Ford* (CVN-78) attained a total cost of US$17.5 billion versus the US$6 billion of its predecessor, the *USS George H.W. Bush* (CVN-77), in great part because of the irresistible urge to replace the traditional hissing and bumping steam catapults with a soundless Electromagnetic Aircraft Launch System that eliminates all that antique steam generation. When the new $4-billion wizardry kept failing, seriously delaying the commissioning of *USS Ford* month after month, the beleaguered

US Navy claimed that the system would save that amount in operating costs over a fifty-year lifespan, indicating that those involved might not do very well as investors. In the Iron Dome project by contrast there was no urge at all to reinvent.

Another procedural factor, which ended up accelerating the project, was the involvement of the different participants very early in the development stage. Usually production work starts only after development is completed, but with Iron Dome there was cooperative synergy with the Ministry of Defense, which became a participant instead of an arm's-length inspecting and auditing customer (a relationship that would have horrified the valiant fighters against waste, fraud, and mismanagement who much prefer adversarial relations without ever computing their tremendous costs). Iron Dome offers the counterexample of a project in which the customer practically merged with the project team. In another drastic departure from all normal practice, the Heyl Avir personnel who were to operate Iron Dome batteries (no way would airmen allow soldiers to shoot missiles into their sky) also participated in the development work.

That the system be easy to use correctly was an extremely important project goal—the batteries would be operated by young conscripts, not seasoned professionals, so air force antiaircraft personnel were brought into the process. Then they surprised the developers by coming up with improvements and adjustments of their own, which were duly implemented. That was only possible because of a most unusual move by the development chiefs: "The [antiaircraft] soldiers who operate the system have our telephone numbers (!) and call us for every problem. [They were] involved in the project from the first moment, from the most basic level of requirements and specifications."[40]

Operating simplicity was a leading requirement, but there were other considerations: "One of the system's requirements that we defined early on was that a small female soldier should be able to step onto the launcher position and activate it. We also had aesthetic design considerations; I told the launcher designer that I want it to look supermodern but also menacing because it is clear that within an hour of its operation, it will appear on CNN and Al Jazeera."[41] Finally, representatives of the manufacturing staff were also integrated into the development process at a much earlier stage than is customary, which lowered development-to-production risks and accelerated problem solving while reducing costs.[42]

A simple timeline captures the sheer dynamics that propelled the Iron Dome project:

February 2006: Gold issues his legal and proper contract for a "technology demonstrator," and no more than that.

August 27, 2006: Minister of Defense Amir Peretz, who happened to be a resident of Sderot, which is very close to Gaza and a prime target of Hamas rockets, declares Iron Dome a "highest-priority project" and calls for an "Emergency Plan" to accelerate its completion. But he does not secure a budget or cabinet support for formal orders though the IDF Chief of Staff.

November 12, 2006: Gold as head of MaFat officially instructs Rafael to initiate full-scale development.

November 16, 2006: the chief of the IDF General Staff Planning Directorate, a one-star brigadier general, assigns responsibility for the development of an "Active Anti-Short Range Rockets Defense System" to the air force, ignoring and cutting across Danny Gold's initiative.

December 1, 2006: Defense Minister Amir Peretz declares that a short-range rocket interception capability is essential, and that Iron Dome is the chosen solution but he cannot himself authorize the needed funds.

February 4, 2007: Prime Minister Ehud Olmert asserts that "Iron Dome is inevitable" and most urgent. But no funding is allocated to Gold's unauthorized venture. Olmert's statement nevertheless encourages the Rafael and Elta managers to keep working confident that all would be put right in the end.

June 4, 2007: the IDF chief of staff, LG Gabi Ashkenazi, withholds his authorization of the Iron Dome project because it still had received no funds from the Ministry of Defense.

July 3, 2007: the new minister of defense, MG (Ret.) Ehud Barak, approves "in principle" Iron Dome's development, which Gold had in fact started two years earlier without any ministerial authorization or funding.

December 23, 2007: five months after Barak's approval, the actual commander in chief under the Israeli system, the Cabinet Committee on National Security that includes the minister of finance, finally endorses Barak's decision, providing initial funding.

January 1, 2008: Iron Dome has its official start—two years and four months from its actual start in 2005.

July 15, 2009: the Iron Dome radar, missile, and software are ready to be tested as an integrated system. It successfully intercepts multiple targets.

July 25, 2010: a prototype demonstrates the selective interception of rockets headed to designated populated places as opposed to uninhabited places.

March 27, 2011: a first operational battery is deployed near the Gaza Strip, followed by a second battery on April 4, 2011, but the Ministry of Defense issues a disclaimer "confessing" that the Iron Dome is not yet fully operational and that more batteries will be pressed into service nonetheless, as an "operational experiment." But that excuse was scarcely called for, as on April 7, 2011, a battery successfully intercepted a 122 mm rocket launched from Gaza toward the city of Ashqelon. The radar instantly identified the launch point, allowing an air force aircraft already airborne on patrol to bomb the launch team, successfully.

August 20, 2011: as the fighting heats up, eleven 122 mm rockets are launched in a single salvo at the city of Be'er Sheva; Iron Dome intercepts nine—and the remaining two cause little damage.

March 2012: a total of some 300 rockets and mortar shells are fired against Israeli territory. Out of 73 rockets identified as real threats to populated places, the Iron Dome intercepts 69. A fourth battery is deployed.

May 18, 2012: duly impressed by its success, the US House of Representatives votes US$680 million for Iron Dome funding, in exchange for technology sharing with US industry, and specifically the major contractor Raytheon.

June 4, 2012: the US Senate Armed Forces Committee approves only US$210 million, still requiring full technology sharing, which does not bother the Israelis because they welcome the added production of interception missiles by Raytheon.

June 23, 2012: Iron Dome achieves its one hundredth intercept.

November 14–21, 2012: Operation Pillar of Defense: Iron Dome batteries intercept 428 rockets for an 84 percent success rate.

November 17, 2012: A fifth battery is deployed in the Tel Aviv area, which intercepts a rocket on the same day.
January 17, 2014: US president Barack Obama approves US$235 million for the procurement of Iron Dome batteries for the United States.
August 1, 2014: the US Congress approves an additional US$225 million to replenish the missile inventory.
August 26, 2014: Operation Protective Edge begins: nine Iron Dome batteries intercept 578 rockets, including larger Fajr-5s and M-75s, for an 89.6 percent total success rate (65 of the intercepts are initiated manually, overriding the Iron Dome's automatic launch program).
May 16, 2016: successful sea trial of the naval version of Iron Dome.
September 17, 2016: an Iron Dome battery intercepts two mortar shells from Syria in the Golan Heights.

Five years later, with renewed hostilities on May 10–21, 2021, Hamas launched a total of 4,360 rockets against Israel. Of these, 1,661 were intercepted by the Iron Dome system, 176 were missed and fell in built-up areas, and 1,843 rockets fell as predicted in open areas.[43] About 680 rockets failed to cross the border and fell inside Gaza, causing significant Palestinian casualties. Ten Israeli civilians lost their lives, with many more saved, some because improvements had increased the system's performance against heavy salvos specifically intended to overwhelm the system.

Strategically, the value of the Iron Dome project was that it offered the Israeli government an alternative to another ground offensive likely to be both costly and inconclusive. The development and early production costs, moreover, were moderate, a total of some $US2.2 billion (half from the United States) from inception to first operational use, in part because of a phenomenally rapid pace of development: the first successful intercept occurred on April 7, 2011, less than six years from the start of the first research efforts in August 2005.

But Michael Lindenstrauss, Israel's state comptroller and keeper of proper government procedures, including safeguards against waste, fraud, and mismanagement, was not amused by the wholesale violation of rules perpetrated by Gold and his confederates. As far as he was concerned, Gold's feat, rather than exemplary, was a case of sustained, piratical insubordination, budgetary misappropriation, and administrative irregularity on the largest scale. The comptroller's report did acknowledge that the circumstances in which Gold made his decisions were far from tranquil and declared his sense of urgency praiseworthy: "The office of the State Comptroller is aware of the decisiveness and fervor of MaFat in producing active defense systems as soon as possible." But it went on to reiterate that it was nonetheless improper for MaFat to initiate full-scale development before the IDF had even defined the operational requirements, before the IDF and the government had approved the spending of vast sums, and before any exploration of offensive alternatives and / or of alternative antirocket defense system configurations. In other words, Gold had picked a system and charged ahead to engineer it into existence, instead of considering all the alternatives. There followed a very long list of irregularities, ending as follows: "Brigadier General Dr. Danny Gold started the development of the Iron Dome in August 2005 in an 'unruly' manner, violating regulations to order the overlapping of the pre-development stage with the full-scale development stage, thereby overriding the exclusive jurisdiction of the IDF Chief of Staff, the Minister of Defense, and the Israeli Government as a whole."[44]

Along with many more violations by Ministry of Defense officials and serving officers energized by Gold's charismatic urgency, the comptroller's report noted that Gold had violated Ministry of Defense decree no. 20.02 of August 2005 when he ordered the "telescopic" scheduling of the project with overlapping stages, so as to commence (unauthorized) full-scale development under the guise of (authorized) prototype development, a violation only possible because of the indulgence of Rafael and Elta management. But parenthetically, the state comptroller noted that the Ministry of Defense had never defined the specific terms it used for R&D projects: Telescopic Development; Spiral Development (seemingly: the sequential reassessment and pursuit of projects); Incremental Development; Specified Demonstration Program; and Technology Demonstration. After the long list of accusations, the report's final conclusions were (unsurprisingly) mild: "It is advisable that the development and procurement of weapon systems (particularly projects which considerably affect the IDF's budget, and it force structure), will be executed after the operational requirements have been specified properly, and approved beforehand."[45]

In all of the above Gold had done what Americans used to do. As late as 1956, when the US Navy started the development of the submarine-launched

Polaris 1A nuclear-armed ballistic missile, there was still room for real project leadership in US defense procurement. Admiral "31-Knot" Arleigh Burke (so called because he would steam his ships above their recommended speed), then chief of naval operations, who combined a fine analytical mind with a hard-charging character, appointed Rear Admiral W. F. Raborn, another brainy, hard-charging character, to head the project.[46] On January 21, 1961, after sixty-six days of submerged patrol, the *USS George Washington*, armed with sixteen Polaris missiles, was declared fully operational—just five years after work had started on a weapons system a hundred times more complicated than the Iron Dome (and funded with a budget a thousand times as large). At a time when there was much anxiety in the United States about Soviet advances in ballistic missiles, the early arrival of Polaris was very important strategically because it was a second-strike weapon, much less vulnerable to surprise attack than bombers on their airfields or Air Force ballistic missiles in their static emplacements on land.

To develop Polaris as quickly as they did, Burke and Raborn had to take many risks large and small, because they were developing an entirely new kind of submarine for an entirely new kind of ballistic missile, whose reduced diameter was made possible by an entirely new kind of nuclear warhead. Polaris only became possible because Burke believed the promise of the eccentric genius Edward Teller, "father of the thermonuclear bomb," that his team could develop a new (W-47) reduced-diameter warhead and do so rapidly. It was on that basis that Burke rejected the sure-thing Army Jupiter medium-range ballistic missile, which would have required much larger submarines, to instead develop the radically different Polaris. But all that happened before the arrival of the present US regulatory regime, which entraps everyone in thousands of rules and an endless sequence of program reviews (and agonizing reappraisals by the buying service) in the name of fighting waste, fraud, and mismanagement. Instead of being honored and praised for bold and successful leadership, these days admirals Burke and Raborn would be hounded out of the service for risk taking, with every imperfection labeled a scandal.[47]

The 2009 report of Israel's state comptroller that both praised and condemned BG Daniel "Danny" Gold shows that Israel too has acquired overlapping authorities that issue contradictory instructions, as well as a great many regulations that assure nothing much while impeding dynamic action.

But in Israel almost daily enemy attacks keep legalistic and bureaucratic degeneration under some control, and the right signals were sent when Gold was not fined or fired but instead honored with the Israel Defense Prize and promoted as head of all research and development in the IDF and Ministry of Defense, with a seat at General Staff meetings.

In 2019 the US Army purchased Rafael's Iron Dome batteries to protect its bases in contested areas. It was a final seal of approval, after ten years of operational activity and the interception of over 2,400 missiles and rockets.[48]

[8]

IDF Women as Innovators

FROM THE START, THE IDF WERE UNIQUE AMONG ALL ARMED FORCES worldwide because the compulsory military service inaugurated with their establishment on May 26, 1948, applied to women as well as men.[1] In the perilous early months of the War of Independence, when many Jewish villages came under attack from their Arab neighbors as well as from roaming bands, both men and women fought in local defense groups, doing their best with their few rifles, pistols, and revolvers. That was hardly unprecedented in human history; women have always fought alongside men in defending towns and cities under siege, with pouring boiling oil (or water more likely) over attackers being something of a literary cliché, while others are depicted sword in hand.[2] In the years just prior to Israeli independence, moreover, Soviet wartime propaganda had made much of the Red Army's fighting females—pilots, snipers, machine gunners, tank crew, and partisans (eighty-nine received the Hero of the Soviet Union award)—who were a small minority of the hundreds of thousands of women in the army, a twentieth or so of its total strength, who were mostly nurses.

The IDF's novelty was that women were not exceptions to be lionized as heroes or relegated to safely female roles (as with World War II British home-service army drivers, for example) but simply conscripts along with the men, who therefore served in a great variety of roles, including combat in the Haganah units that transitioned into the IDF ground forces and in the elite

Palmach striking forces. As more organized IDF forces emerged, however, women soldiers and officers, while still trained to use rifles and pistols, were assigned to noncombat roles as radio operators, headquarters staff, office secretaries, storekeepers, bookkeepers, and military nurses.

Conscription was also different because women served shorter terms. Initially service was twenty-four months for both genders, but gradually men's service was lengthened to thirty, then thirty-six months, and then shortened again to thirty-two months. Another difference is that a significant proportion of women were legally exempted, including females married by age eighteen, not uncommon among some socioethnic groups, and those who applied for exemption on ideological grounds and could meet set conditions: a convincing declaration that the applicant cannot serve because of reasons of conscience or because of a "religious way of life," which is accepted if the applicant "keeps the laws of Kashrut at home and outside and does not travel on the Shabbat," these being observable practices. Yet another difference that lasted until 2001 was that female soldiers and officers remained under the managerial and disciplinary authority of the Women's Corps (no matter in what unit they served). The corps was in charge of induction, recruit training, and transfers among different IDF units; it also operated soldier-teacher units, which augmented local teachers in new immigrant townships and remote communities.

The post-1949 regime changed radically with the expansion of the field forces after the 1973 war, in which IDF frontal forces were badly outnumbered by expanded Arab field armies. Personnel therefore had to be found for many more frontline combat units by slimming down everything else—service and logistic units, supporting commands, headquarters, and, most damagingly, the instructor cadre of the training schools large and small. Even before then, Colonel Avishai Katz, commander of the Military Engineering School (acronym Bahalatz) Ba'had 14, had started a new program in 1972 that enrolled women as combat instructors. Its evident success spread the practice to the infantry and armor-training bases, and then throughout the IDF.[3] A new policy emerged to employ selected women as trained instructors in all areas, notably including weapon and armored-vehicle training.[4] IDF male recruits learn from young women aged eighteen to twenty specialist and technical skills from sniping to the emplacement of demolition charges, operation of all field radios and assorted sensors, artillery gunnery, the skills needed by armored fighting vehicle crews, and more. Started because of harsh necessity in the 1970s, when Israel did not have enough male conscripts both to staff its field units and to serve as instructors in all its training courses (any intensively trained military force needs many instructors), the practice was institutionalized when the IDF discovered that women often made better instructors than men, not least because they could more easily strike the right balance between discipline and sensitivity in dealing with young conscripts.

Because their own training is very thorough and employs highly effective pedagogic techniques, women instructors earn the respect of their male colleagues in the IDF, and the close attention of their male pupils.[5] Young recruits in their first encounter with dangerous weapons, such as hand grenades, are reassured by the familiar ease of the women instructors who do it first. Aspiring gunners of the huge, high-velocity 120 mm cannon of the standard Merkava tank may see their instructor next to them inside the tight tank turret nonchalantly flicking away a strand of stray hair as the gun's three tons of steel recoil back explosively two inches from their right ears.[6]

The IDF's training needs are certainly unique: among all the armed forces in the world it alone must rely on teenaged conscripts to operate even the most complex weapons and support systems, instead of the professional, technically trained or long-service noncommissioned officers of every other armed force.[7] Therefore, even before the arrival of the women individually tested and selected for instructor training, IDF pedagogic methods were very carefully considered, in order to capture and retain the attention of young soldiers and teach them what they have to know. The advent of female trainers added an element of gender tension to the motivating force of the established pedagogy, as young male recruits would strive to the maximum to avoid failing in front of their women instructors.[8]

The Women's Corps, which had affirmed the separate and different role of women, became increasingly obsolete as women served as trainers in all parts of the IDF, with some going beyond in combat support and then combat roles. Roni Zuckerman, the first female jet-fighter pilot, received her wings in 2001. That same year, after years of declining importance, the Women's Corps was finally abolished and replaced with the radically different office of the Women's Affairs Advisor to the Chief of Staff, charged with enhancing the role of women in all capacities by ensuring more opportunities, promoting suitable unit environments for women soldiers, and assimilating women into military leadership positions at all ranks. Soon individual women volunteered for all manner of roles in the ground, air, and naval forces, if not in direct combat except for dedicated units, near enough to combat in combat support roles from artillery to airborne search and rescue. Some women instructors exploited opportunities to go into combat—thus, a driving instructor for a heavy armored personnel carrier (APC) joined an operation in Gaza when the APCs in her training unit were allocated to equip an infantry unit that had none of its own. Instead of sending women soldiers en masse into existing combat units in pursuit of formal equality regardless of the practicalities—the IDF lives rough, in tents or huts, with outdoor showers—specially structured combat units of female and male soldiers were established to provide needed facilities, and also to accommodate upper-body strength differentials (attempts to deny physiology have caused high injury rates among women soldiers in other armies).

In 1995 the Border Guard, administratively under the police, but operationally often under IDF command, opened combat roles for women conscripts, both as riot police and as counterterrorist light infantry. Following the Border Guard's successful integration of women combatants, a mixed combat unit was formed in 2000: the Caracal Thirty-Third Border Security Battalion, more than 50 percent female in 2021. Its soldiers are trained to patrol the borders for armed infiltrators and smugglers of drugs and humans. By taking on the border security role, the Caracal and other mixed-gender infantry battalions allow the first-line infantry units (e.g., Givati, Golani) more training time.

The IDF leadership, including the Women's Affairs Advisor to the Chief of Staff, had resolved that if the IDF were to have female combat soldiers, they would have to be really good soldiers, and Caracal as the first mixed combat unit therefore had to be tested not in the safest possible sector but the very opposite. Indeed Caracal's women and men soon had to fight off well-armed ISIS infiltrators from the Sinai. In one incident Captain Or Ben Yehoda led her company against some well-armed infiltrators; she was injured but her force was able to kill six of them. A female officer and a female sniper were awarded citations for their bravery and performance during similar fights.

Soon recognized as a success, Caracal became the model for two additional mixed light-infantry battalions: Lions of Jordan in 2014 and Cheetah Battalion in 2015.[9] Many other women have served in individual combat roles in artillery units, in antiaircraft units, in transport, and as combat pilots, navigators, and officers in every branch. Orna Barbivai, the first female major general *(Aluf)* in the IDF—one of very few major generals in the forces—was promoted to that rank in June 2011 when she was appointed head of the General Staff personnel department.

The compulsory service of women in the IDF was a unique innovation at the start in 1948 precisely because it was neither a short-term remedy for the manpower shortage nor a propaganda stunt, but rather a case of making the absolute best use of scarce human resources. In a country where almost all eligible men but for the most religious serve in uniform—as near a universal military service as one can find anywhere in the world—many women did not serve because they were the daughters of traditionalist families, in which military service was seen as subversive of the modesty expected of all young women. One unintended consequence was that the IDF became the emancipator of a great many women who rebelled against noxious old-country customs by running away from home to recruitment bases to join the secular women of their age cohort. For the same reason, some Arab women, mostly Christian, also volunteer for service in the IDF, some serving in combat roles.[10]

[9]

Military Doctrine and Innovation

WHEN THE IDF HAD THEIR BEGINNINGS IN 1948 UNDER THE DIRECTION of the country's first prime minister and minister of defense, David Ben-Gurion, the British Army was both the detested ex-occupier that had striven to disarm the Jews as they faced deadly attacks, and Ben-Gurion's favored model of an apolitical army of the entire nation. Under British rule the Jews had gradually built up not an underground army but rather, politicized militias. The Haganah was by far the largest, controlled by Ben-Gurion's Mapai social democratic party.[1] Its war-winning elite force, the Palmach, was mostly led by members of the more left-wing Achdut Haavoda party, including future prime minister Yitzhak Rabin and the outstanding field commander of the War of Independence, Yigal Allon.[2] Their ideological rival, the Irgun, was led by the right-wing "Revisionist" party.[3]

Ben-Gurion could in theory have made himself a dictator, given that his party's militia controlled most Jewish areas of the country.[4] Nevertheless, for Ben-Gurion political control of the Haganah represented not an advantage to be exploited but rather a dangerous confusion of roles—he was, relentlessly, a statist among ideologues. As minister of defense as well as prime minister he therefore promoted those who had volunteered to join the wartime British Army rather than the recently victorious commanders of the Palmach, "a political youth movement in arms" that despised military formalities and even proper uniforms, whose pre-1948 inspiration was the Soviet-directed partisans who had fought behind German lines. Ben-Gurion by contrast wanted regular forces of disciplined, uniformed soldiers led by professional officers, just as in the British Army.[5]

But when it came to fighting methods, particularly for the ground forces, not even Ben-Gurion was impressed by the decidedly undynamic British style of war, which relied on vastly superior firepower—artillery barrages and air bombardment—to defeat the enemy by sheer grinding attrition, before step-by-step advances of armor and infantry to seize the ground won by firepower. This method was entirely useless for the Israelis because it required greatly superior forces, with three-to-one or greater numerical advantages in men and firepower, just as at the Battle of El Alamein, and indeed in almost every British victory against the Germans. Therefore, while the IDF were to copy the nonpolitical ways and organizational ideas of the British Army, they could not possibly copy its fighting style and operational methods, because they had to be able to win as they had won in 1948, even when inferior in numbers and firepower. Only agile maneuver warfare could circumvent, infiltrate, dislocate, confuse, and disrupt greatly superior enemy forces by bold surprise actions or by acting and reacting faster than the enemy could. Obviously, this required fast-thinking commanders willing to take risks—the kind of people who lead innovation in peacetime.

The Palmach commanders were the right ones for the task; their fast and fluid tactics, their bold operational methods in the War of Independence, exemplified maneuver warfare at its best.[6] Palmach units had won their battles in 1948–1949 with some heroic, and very costly, hand-to-hand fighting (some by fifteen-year-olds), but mostly with bold, fast-moving offensives that outmaneuvered Arab forces that were larger and better equipped at first, but were fatally slowed by rigid top-down chains of command.[7] In other words, the Palmach's style of war was pitched at the "commando" end of maneuver warfare, reliant more on speed and surprise than firepower or mass—and that remained so even when the combined forces from three separate Palmach brigades could mount large operations in the closing months of the war together with regular IDF units.

Although the Palmach's outstanding field commander Yigal Allon, who had led every major campaign of the war, left the IDF after the war to enter politics, other Palmach officers remained in the IDF to propagate their ethos and methods. Also, there were other officers who did not belong to the Palmach but nevertheless favored its methods, most notably Moshe Dayan, an exemplar of the fast-moving style of war in 1948 as commander of a jeep battalion, then in the 1956 Sinai Campaign as chief of staff and the 1967 war as minister of defense. In this early phase of the IDF's emergence, one inspiration was the fast-moving cavalry of the Red Army at its best during the later phases of the Russian Civil War, communicated to the Palmach by its most senior figure, Yitzhak Sadeh, a decorated veteran and company commander in the Imperial Russian Army.[8] Also much admired in the early postwar years was Soviet partisan warfare in German-occupied territories, fought by assorted volunteers under the leadership of Soviet Army officers. But only a handful of Jewish partisans survived to arrive in Israel and describe it in any detail, so that it mostly influenced the Palmach's marching songs and its nonuniforms, including the use of knitted woolen socks as headgear. But most important was the notion that sheer speed in decision and action can outweigh mass; fundamental for the Palmach, it remained fundamental for the IDF.

Another source of the Palmach's military culture was the theory and practice of the brilliant eccentric and subsequent field commander Orde Charles Wingate, the rarest of birds as a pro-Jewish British officer in Palestine.[9] Forming a temporary mixed British-Jewish unit for counterguerrilla operations in 1938, he taught his disciples, most notably Yigal Allon and Moshe Dayan, that enemies are best defeated by sudden raids, with all-important surprise gained by hard marching at night on unexpected paths or by ambushes that achieve surprise by stealthy positioning and stoic patience. Wingate's formula could not be applied by doggedly obedient, let alone reluctant, soldiers. It required highly motivated, well-trained, and physically fit fighters, but not many of them, because with surprise even small numbers could win. From this calculation, backed up by the acceptance of strenuous training exercises, still perpetuated in the IDF's long marches, came the commando element in Palmach operations. This in turn inspired the IDF's own raiding culture, begun in a small way with the single Unit 101 but developed over the years into a spectrum of different commando units each specialized in one task or several. Above all, in the IDF the commando element—maneuver warfare at its maximum—is not peripheral as in other armies, because many senior officers are promoted from the commando units.

Thus, two radically different military doctrines were influential in the making of the IDF: the systematic, sometimes ponderous British doctrine that minimized risks but also gains; and the high-risk / high-payoff Palmach style of maneuver warfare that tried to exploit surprise to defeat many with few.

It follows that when outsiders suggested the adoption of some new tactical or operational method, their suggestions could not be rejected out of hand, because they collided with two official doctrines that contradicted each other.

As soon the War of Independence ended with the 1949 armistice, the IDF tried to upgrade the professional education of its still-young senior officers by sending them to study in European war colleges or in universities on occasion. Thus Moshe Dayan, future chief of staff and minister of defense, attended the British Army's three-month Senior Officers' School in Devizes in 1952, after he had already commanded forces in war, negotiated an armistice, and headed an area command. Though the personal soldier-servant ("batman") who woke him up each morning with tea and polished his shoes amazed Dayan, he did rather like the way problems were set and solved by the battle-experienced instructors.[10] Conversely, Field Marshal Bernard Law Montgomery of El Alamein would sometimes visit Devizes to teach the "Montgomery method" of winning battles, starting with the accumulation of vastly superior artillery, armor, and infantry forces—not very useful for IDF officers who had to learn to fight outnumbered and win.

Other IDF officers sent to British or French staff schools and war colleges were grateful for the diversion at a time when foreign travel was an unattainable luxury for most Israelis, but they reported learning very little that was useful—the suggested tactics and operational methods were much too rigid and bureaucratic for the improvisational Israelis, and presumed weapons and firepower far beyond Israel's means. Ezer Weizman, already a successful combat pilot and future air commander and president, was sent in 1951 to the Royal Air Force Staff College at RAF Andover, where he too concluded that it was important not to learn from others.[11]

But the young Israeli commanders did learn planning methods, logistic calculations, and orderly staff-work procedures useful for the often too informal, too improvisational IDF. Nor could extreme antipathy completely deny the obvious virtues of that other model, the German style of fast and fluid maneuver warfare, propelled by bottom-up leadership from the front, in turn made possible by the Prussian General Staff's invention of supervising but noninterfering higher-echelon headquarters, whose staff officers were only to step in to coordinate separate forces advancing under their own opportunistic leaders if they were about to collide or when they could converge against the enemy. At the time, the most recent expression of that style of warfare was the blitzkrieg campaigns of 1939–1943, in which columns of trucked infantry spearheaded by tanks would advance as rapidly as possible, bypassing any strong local resistance instead of stopping to fight it, unless the enemy could quickly be dispersed by artillery fire or highly focused air attacks.

Although most of the German Army of the Second World War consisted of foot infantry and horse-drawn artillery rather than armored forces, while the bombing capacity of the Luftwaffe was not large by later Anglo-American standards, the propaganda of the time shows columns of tanks rushing forward and waves of Stuka dive-bombers descending on targets. That misrepresentation was actually the very essence of the blitzkrieg, in which the enemy was to be shocked into retreating pell-mell, with panic making up for the German Army's lack of much armor or firepower. (Once the Russians acquired both in vast amounts, the ruse was over.) There were important lessons in this for the IDF, which had to learn to both exploit psychological effects, and also not to rely on them overmuch.

Another dimension to German-style warfare long predated the Nazis and blitzkrieg: the careful cultivation of tactical skills in serious training courses, which was the German Army's most enduring strength, persisting even when firepower superiority and all else were lost. Even in the last weeks of the Second World War, a German infantry unit holding a position that still had some experienced noncommissioned officers and enough ammunition for its machine guns was an immovable object even with massive firepower superiority. More specifically, German tactics had the effect of turning the defensive into something resembling the offensive by seeking every opportunity to counterattack, even if only with a handful of soldiers; on the offensive, on the other hand, most important was to find a way to reach behind the enemy force instead of driving it back with frontal attacks. That was the offense / defense doctrine the IDF needed to capitalize on their bright soldiers and the training intensity possible with prolonged military service.

In 1924, long before the IDF were born, three Haganah members traveled to Germany to be personally trained by Paul Emil von Lettow Vorbeck, the extraordinary German commander who had been trained as a regular Prussian military officer for warfare in Europe, but then achieved impossible guerrilla victories as a commander in East Africa.[12] With German East Africa blockaded from August 1914, von Lettow received no further supplies or reinforcements. Surrender to the superior British forces coming up from South Africa was the only realistic option, but von Lettow advanced, retreated, and counterattacked again many times during the next four and a half years, surrendering only on November 23, 1918, two weeks after the official German surrender. With a handful of German officers and the African troops he often trained personally throughout the war, von Lettow had successfully waged a war of relentless maneuver, supplying himself by infinite improvisations, including the production of his own gunpowder.

Improvisation, not least coming up with equipment and supplies that would not just be issued, was one art von Lettow could teach in spades, along with the central and unusual idea that the offensive was the only proper stance for an inferior force, attributes that both the prestate Haganah and the early IDF would desperately need. And the 1924 visitors could also learn much from the Reichswehr, the postarmistice German army limited by treaty to 100,000 personnel in all (with another 15,000 for the navy) and severely restricted in equipment (e.g., no tanks). It had to make much from little, notably by training every soldier to the level of an instructor to make feasible a rapid mobilization of civilians.[13]

The Haganah's own officers' course began in 1937 under the entirely self-taught Yosef Avidar, who relied on military journals to keep abreast of European military thinking.[14] He envisioned a small army of well-trained leaders, similar to the Reichswehr, in which every soldier would be trained at least to the level of a sergeant. Avidar held that the German texts were best, but he also relied on Soviet, British, and Polish publications.[15] Many future leaders of the IDF were Avidar's trainees, including both Moshe Dayan and Yigal Allon. And though Haganah units were very small, Avidar trained his pupils to command battalions and even brigades, which was fortunate because in the 1948 fighting, the newborn IDF went from fighting by platoons of thirty to fighting by combined brigades of thousands in just a few months.[16]

In structuring the IDF's general staff, their high command, Ben-Gurion's model was again British, but he was also influenced by the recommendations of another foreign influence, the newly retired US Army colonel Fred Harris-Grunich, who argued for example that the intelligence branch should be separated from the general staff, as in the US Army.[17] Another influence on the structure of the IDF was the impact of wartime volunteers in the British forces, including those who had become pilots in the Royal Air Force. They pressed for an entirely independent force like the RAF, which was the first air force to become a separate service at a time when armies and navies elsewhere each had their own air branch, and there was no independent air force anywhere. That demand was rejected but as a compromise, the Heyl Avir, or Air Corps, was granted a higher status than the artillery or armor because it was allowed to have its own command headquarters, though it remained subordinate to the General Staff. A similar solution was found for the Heyl Hayam, the IDF's Sea Corps.[18]

Everything had to be done on a shoestring and in a great hurry, but the officers who designed the IDF piece by piece were well informed about the organization of the general staffs of Western forces, and tried to pick and choose among their components to design a headquarters organization that would suit the IDF's special circumstances, including the need to rely heavily on reserves and address extreme scarcities. In September 1947, six months before the establishment of the IDF, a British Army veteran, Major (Ret.) Haim Laskov, set out to translate more than thirty British training manuals, and wrote out course programs based on them.[19] What could not be learned from the British Army was learned from other armies, notably the Swiss and Finnish for their reserve army model, to which female conscription was added in response to the extreme shortage of fit personnel.

Eventually when some IDF officers visited the US and French armies in the 1950s, they were awed by their magnitude and traditions but unimpressed by their methods.[20] When the IDF started to develop armored forces, interest in German armor operations focused on their methods and tactics in the Western Desert, which resembles the Negev and Sinai in both terrain and climate. Uri Ben Ari, one of the founders of the IDF's Armored Corps, was born in Berlin and was a native German speaker. He read the original German armored warfare manuals and conducted various trials to try to determine if they suited IDF needs.[21] It was mainly because of his work that the IDF adopted, among other things, armored tactics based on German rather than British combat doctrines.[22] Following the establishment of diplomatic relations with West Germany in 1965, the IDF even sent—not without hesitation—officers to study at the German staff college.[23] The rapid advance of the IDF's armored forces into the Sinai in the 1967 war certainly evoked memories of the German blitzkrieg.[24]

As for naval doctrine, Israel needed something rather unusual, because it had no combat vessels at all nor any prospect of acquiring any, while Egypt had several. As noted earlier, largely because of Yohai Ben-Nun, an answer was found in Italy's highly successful Second World War combat frogmen, who used newly invented masks, fins, and air tanks of amateur divers, and their own inventions of manned torpedoes, explosive boats, and magnetic limpet mines, to sink several major British warships. Ben-Nun started the Shayetet ("Flotilla") 13 sea commando force, for which he brought Italian war-surplus diving and frogman equipment, along with Italian expert Fiorenzo Capriotti, a veteran of the famed and notorious Decima MAS (whose land troops were guilty of many atrocities) with the approval of Ben-Gurion.[25]

Capriotti started training his Israeli pupils with the simplest technique: explosive boats, where the trick was to keep steering at the target vessel, only jumping off at the last minute. Before the training was complete, the pupils put their skills to the test on October 22, 1948, when the Egyptian flagship *Amir Farouq* and an escorting vessel were detected sailing off the Gaza shore. Ben-Nun and three other Capriotti pupils went into action, sinking the *Amir Farouq* with its 500 crew members, and severely damaging the escort vessel. Capriotti had very much wanted to join in, but his pupils refused, to his great frustration.[26] Ben-Nun, who drove the boat, was to continue as Israel's naval innovator by conceptualizing a navy consisting of small but highly capable missile boats armed with the Gabriel missile he initiated.

Thus, with British military traditions, German maneuver warfare, Italian asymmetrical naval warfare, Wingate's commando methods, and homegrown variations of all of the above, the IDF did not start off with a coherent military doctrine—and it would never acquire one. Nor did the leading minds of the IDF ever subscribe to the idea that long-term planning, based on forecasts of the strategic environment ten years out and more, could guide the development of the IDF. In the unending turmoil of the Middle East, even five-year plans are soon overtaken by events, so that any multiyear programs mandated by the mechanics of procurement soon require changes and adaptations—the IDF slogan that "plans are merely a basis for changes" is certainly validated by the record. In 2003 a five-year force-building plan was initiated; after the 2006 Second Lebanon War most of it was scrapped to pursue a completely different focus, but that plan was stopped halfway because of funding problems. Following a period of indecision, a new plan was formed that also incurred financing problems just being remedied when the 2011 Arab upheavals took away the Syrian military threat and largely changed the Arabian scene. That in turn prompted a new plan resembling the 2003 plan, until the Gaza fighting in the summer of 2014 revealed the plan's limitations, eventually prompting an all-new Gideon plan of radical innovations launched in July 2015, followed in 2019 with yet another new plan, Tnufa, to exploit emerging technological capabilities.[27] Without a coherent military doctrine, and without force-building plans that last long enough to matter, there is no controlling set of ideas, an absence that leaves some uneasy but certainly leaves the door wide open for new ideas in the IDF—ideas from anywhere, which certainly favors innovation.[28]

[10]

From Triumph to Failure in the Air, 1967 and 1973

WHEN AMBITIONS EXCEED AVAILABLE MEANS, SENSIBLE PEOPLE TRIM them back, but in the 1960s the Israelis, facing the rapid, Soviet-supplied growth of the Arab military forces around them, could not afford to be sensible. Instead of accepting their limitations, they plunged into attempts to find out-of-the-box solutions, preferring the risk of failure to the meek acceptance of unhappy realities. David Ben-Gurion could never have agreed with Otto von Bismarck that "politics is the art of the possible" when he set out to build a state without an army to defend it—an army only a state could build. Israelis in uniform, from private soldiers to major generals, are just like other Israelis in that regard, only more so, and routinely prefer risk-taking to settling for the unacceptable.

Back in 1956, when the Heyl Avir still operated piston-engine P-51 Mustang fighters to supplement its meager total of sixty assorted jets, a new mission was found for that famously versatile aircraft on the first day of the Sinai Campaign, October 29.[1] It would disrupt Egyptian communications by cutting the phone lines strung from pole to pole across the peninsula. Four Mustangs were specially equipped for the task by attaching wires to their tails weighted down to form hooks in flight.[2] But that device failed when the wires simply fell off. Instead of giving up, the pilots cut the telephone lines with their propellers and wings, adding more risk to an already chancy mission, flying so close to the ground.

For the airmen, the only possible response to the lack of means—ensured by the persistent US refusal to sell combat aircraft to Israel—was to find a way to do much more with less.[3] In the same 1956 Sinai Campaign, the 103rd "Flying Elephants" squadron that grouped Israel's tiny inventory of transport aircraft delivered 351 paratroopers at the Mitla Pass in the Western Sinai in the opening move of the campaign, a rare combat jump deep in enemy-held territory.[4] (The copilot of the lead aircraft was Israel's first female combat pilot, Yael Rom.) There were so few transports that of the ten C-47 Dakotas, the military version of the ubiquitous DC-3, seven had to be borrowed at the last minute from the French, then Israel's all-purpose ally.

It was with this adventurous background that the air force of the 1960s approached the seemingly impossible goal of operation Moked (Focus): to destroy the totality of enemy airpower at the very start, so as to nullify Arab numerical superiority in the air, prevent bombing of Israel's civilian rear, and allow the use of surviving Israeli combat aircraft to support the ground forces. It was an undertaking that required the exact opposite of Israeli-style dash-ahead-and-hope-for-the-best improvisation, calling instead for a relentlessly systematic, all-encompassing effort to overcome every impediment to an air offensive meant to destroy the required number of aircraft in each of many different airfields within the shortest possible time—and to do so with an air force simply too small for the task. For the improvisational Israelis it required a veritable cultural change to attain an operating discipline intense enough to achieve extreme levels of performance every time. To relentlessly train to precise standards, with no tolerance for deviations, was alien to the prevailing Israeli mentality. But it was done.

When the Moked plans were finally refined just in time for the war that started on June 5, 1967, Israel could count on a grand total of 203 strike aircraft of all types.[5] Their payloads ranged from a mere 500 kilos to a maximum of 4,000 kilos over the required combat ranges, with the mainstay Mirage IIICJ limited to 3,000 kilos.[6] That arithmetic generated a combined total weapon-load per sortie equivalent to that of just five B-52 bombers—a relevant comparison because Moked was essentially a bombardment plan, in which air combat would only take place if things went badly wrong, an eventuality for which no proper provision could be made because there were simply not enough aircraft for a serious air defense as well. Forty-four Fouga Magister training aircraft armed with ten rockets were used to support ground forces, while the strike aircraft were focused on Moked.

The IDF was badly outnumbered compared to the combined Egyptian, Jordanian, and Syrian air forces (544 combat aircraft versus Israel's 203), and more so when Iraq with another 100 or so aircraft intervened as well.[7] But this was less of a problem for Moked than the sheer number of air bases that had to be attacked simultaneously. Egypt alone had a total of eighteen: four just across the border in Sinai, three farther back along the Suez Canal, six in the Nile Delta above Cairo, and five deeper in Egypt, most importantly the Beni Suef base holding thirty Tupolev-16 (Tu-16) medium bombers, the strongest strike force in the Middle East with a combined payload of 270 metric tons, larger by itself than Israel's total.[8] In addition, the Syrians, who were certain to go to war, operated five air bases, Jordan had two bases, and Iraq had one within fighter-bomber range of Israel near the H-3 pipeline pumping station.[9] Thus during the first day of a future war, the Israeli Air Force would have to attack a total of twenty-six air bases in four different countries.

The plan called for an all-out strike: on June 5, when the fighter-bombers went out to an uncertain fate, only twelve Mirages were left behind on air defense alert, to serve as a desperate last-stand force in case Moked failed (four were sent up with eight others waiting on the flight line, ready to join the subsequent waves of bombing, or else take off, jettison bombs, and engage incoming enemy fighters if necessary).[10] Fifteen lightly armed training aircraft, Fouga Magisters with rockets, were employed to beef up the initial strike force—they would strike a number of less defended Egyptian radars.[11] After the first wave, surprise would be lost, and the second wave would face an alert enemy. It was evident that second-wave aircraft flying in loaded with bombs would therefore be desperately vulnerable to any unencumbered fighter flown by any passably competent pilot. So maximizing the number of bases attacked in the first wave and choosing which bases to strike in the first wave was critical.

Maximizing the number of bases required attacking with fewer aircraft per base. This meant fewer Egyptian aircraft hit per strike, thus allowing others to take off; so all the Egyptian aircraft on these bases had to be pinned to the ground to prevent them taking off both during the first wave and again later in the second wave. The straightforward remedy was to bomb all the runways of each attacked base in the first passes of the first wave. That is why Intelligence was told to make great efforts to ascertain the exact nature of the different runways—their vulnerability to bombs is a function of the nature and quality of their construction. But the obvious problem with that obvious remedy was that cement runways are tricky targets for anything but steep dive-bombing with heavy bombs, because in shallower dives the bombs are apt to slither away, while small bombs may just pockmark runways in ways easily patched up; by 1967 quick-drying cement fillers and light-alloy slats were standard issue for all air forces.

Hence the remedy was no cure because with so many runways to attack, allocating enough weapons to incapacitate them, even at a minimum rate of eight 500-pound bombs for each one, would take some 60,000 pounds, drastically diminishing the number of targets that could be attacked in the first wave. The solution was the one genuine technological innovation of Moked, invented by Israeli engineers only in 1966: an entirely new kind of runway-busting bomb that weighed only seventy kilos and thus had a very low cost in payload, but could still blast out a crater of as much as 5 meters in diameter and as much as 1.5 meters deep—too deep to be usefully patched. This PaPaM bomb (Ptsatsa Poretset Masloolim—"Runway Penetrating Bomb") overcame the limitations of its small explosive power because it was first dropped in level flight at the low altitude of 300 feet (100 meters); its forward momentum was then halted by a small parachute brake, whereupon it would tilt down toward the runway with gravity, at which point a small rocket would ignite to propel the bomb deep into the ground, even through concrete, with the explosion delayed for six seconds for maximum damage.[12] It all seems much too complicated to work but it cratered runways beyond quick repair.

Another experimental munition, informally known as Olar Khad ("Sharp Pocketknife") was bigger than PaPaM, but was not operational in time for Moked.[13] It was destined to have a curious afterlife because it was persistently misidentified even by reputable historians as the much bigger French runway-buster Matra Durandal (BLU 107 in US service), which was indeed very effective. But its crippling shortcoming for the Moked planners was that it did not yet exist in 1967, only entering service ten years later.[14]

It was decided to attack the parked planes only with cannon, so that the entire limited bombload could be dedicated to striking the runways and ensuring they were hermetically blocked, enabling more bases to be attacked simultaneously. Aircraft not destroyed in the initial wave would not be able to take off and could be dealt with later. Every Israeli fighter, except the Ouragans, mounted a pair of 30 mm cannon, the formidable French DEFA 552. Each carried 125 powerful 30 mm rounds (the British had the equally formidable ADEN gun; both were copies of the German Mauser MG 213C, the first revolver air cannon). Even a single 30 mm hit on target could disable an aircraft, and a second one could destroy it, so that scant bombloads could be powerfully complemented by strafing attacks. The Ouragans mounted four 20 mm cannon each, powerful enough even if less lethal than the DEFA 552s. Although the cannon were French, as were the fighters that mounted them, in the case of the Mirage IIIs there would have been no cannon but for the obdurate insistence of the Israelis, because the French (like the Americans and British at the time) believed cannon to be obsolete for ultramodern fighters armed with missiles.

On the Israeli side, as noted earlier, the decision-makers were fighter pilots, not engineers. In their exercises they had learned that supersonic fighters could not engage each other at supersonic speeds, because human pilots could not see their targets and guide their aircraft against them at those speeds. Existing missiles were not effective enough to compensate for these deficiencies. It followed that most air duels would still be fought at shorter ranges, allowing the use of cannon, which were also very much cheaper per shot, as well as useful and indeed necessary to strafe targets on the ground. In 1967 the Israelis' insistence on having cannon on all their aircraft paid off strategically. However, the first lesson for all air forces was to build heavily protected hangars for their aircraft so in the future only special bombs could penetrate and destroy the aircraft within.

To catch as many Egyptian aircraft as possible on the ground in the initial attack, the exact timing of H-hour, when the first bomb was to be dropped, dictated the timing of every other action that morning. The common military wisdom is that H-hour should be either at dawn or at dusk, when aircraft approaching targets would be less visible. Both options were considered, but air intelligence officer Yeshayahu Bareket did not agree, because everybody knew this theory and therefore the Arab air forces were routinely alert at this hour with some planes in the air and others ready for takeoff. He gathered his intelligence analysts, mostly young conscripts nineteen to twenty-one years old, for their opinions. Based on detailed knowledge of the daily routine of the Egyptian Air Force, a nineteen-year-old came up with the idea of an 08:00 H-hour, because it was then that Egyptian pilots on flight duty traditionally took a break to eat breakfast after their routine dawn alert and early-morning training flight. Bareket managed to convince Moti Hod, the air force commander, thereby prevailing over the majority view at headquarters. Thus, the single most important planning decision, the H-hour, was suggested by a nineteen-year-old corporal.[15]

The attacking aircraft took off in sequence on June 5, 1967, from 07:10, each timed to arrive over its target air base (including distant Luxor) at the H-hour of 07:45. To maximize surprise, the takeoff sequencing and flight routes were planned to ensure a simultaneous H-hour at each Egyptian base. In the first wave, ten Israeli aircraft were lost to antiaircraft fire; nine Egyptian aircraft that were on patrol, or managed to take off, were shot down in air combat; and a total variously estimated at 197 to 204 were destroyed on the ground.

Moked's second wave started at 09:34 the same day, when some 164 air force fighter-bombers that had returned from the initial attack, refueled, and rearmed flew out a second time to attack sixteen Egyptian air bases, some new targets, and others previously attacked but insufficiently damaged. A third wave, beginning at 12:15, completed the mission, destroying another 107 aircraft on the ground, thus reaching a total of 310 Egyptian aircraft, of which 286 were combat aircraft destroyed by the first two waves, in a total of some three hours.[16]

The third and fourth waves had planes available for new targets. Israel planned to fight only Egypt, but the Jordanians, Syrians, and Iraqis had been summoned to join the war by the Egyptian high command, which claimed both that it had destroyed much of the Israeli Air Force and that its ground forces were advancing to Tel Aviv. Jordan duly opened artillery fire on Jerusalem and the nearby border villages, with the only long-range guns in the Middle East, US-supplied 155 mm "Long Toms" firing deep into Israel. Sixteen Hawker Hunter fighter-bombers of the Jordanian Air Force bombed civilian and military targets in Israel. Concurrently, Syrian artillery up on the Golan Heights fired down into the Hula Valley, and twelve Syrian MiG-21s bombed civilian and military targets, as did three Iraqi Hawker Hunters and a single Iraqi Tu-16 medium bomber.

The damage the Jordanians, Syrians, and Iraqis inflicted on Israel was strategically insignificant, although twenty civilians and a dozen soldiers died and hundreds were wounded, but the IDF's response was decisive. By 12:45, five hours into the war, eight Israeli fighter-bombers were sent to attack the two Jordanian air bases, destroying all their aircraft. An additional eighty-two sorties flown against Syrian air bases destroyed some sixty aircraft.[17] There were also long-range attacks against Iraq's H-3 forward air base, where ten aircraft were destroyed, and against Egypt's most remote air base at Ras Banas far down the Red Sea, completely out of range for all Israeli combat aircraft except for its Vautour light bombers.

By 18:00 June 5, the total number of Israeli strike aircraft that had gone into action that day was almost the same as the total number of strike aircraft in squadron service, which in turn approximated the totality of the air force active inventory, a phenomenal availability / serviceability rate of 100 percent, or near enough, especially noteworthy for an air force forced to keep odd lots of older fighters in first-line service to make up numbers. Twenty had been lost, as well as two rocket-armed training aircraft supporting ground troops and one transport aircraft hit on the ground by Jordanian aircraft. Normally, well-run air forces around the world are content with 50 percent availability rates. While everyone understands that what counts in war is not the number of aircraft in inventory but rather the number actually ready for action, such aircraft readiness that requires much effort and much expense is as perishable as cut flowers that must be bought anew each day. To have aircraft fully up with all systems and subsystems working perfectly requires both thorough maintenance work after each flight and a large inventory of costly replacement parts. Thus, the more readiness an air force buys, the less money it has to buy everything else, notably more aircraft.

What Israel needed, however, was both a relatively large air force for a country with its poor economy and also high availability rates. One way to resolve the contradiction was to invest in the local production of replacement parts. Israel's aircraft industry was then very small, by no means advanced and not especially efficient, but the Dassault company of France—then Israel's only supplier of combat aircraft—was almost as famous for the exorbitant prices of its replacement parts as for the design talents of its engineers. Hence even if it had to function on a very small scale, local production could be cheaper.

Accordingly, anything that could be machined locally usually was, incidentally imparting production skills that would soon become useful in producing the semibootleg Nesher copy of the Mirage once France rewarded Israel for its 1967 victory by imposing ever-tighter restrictions on further arms sales, culminating in a full embargo in January 1969.[18] Anything that could be saved from the annual replacement-part budget could be rolled back into buying more aircraft or at least more ordnance, as well as fuel and consumables, everything from batteries to high-wear components.

But the principal out-of-the-box remedy for the impossibly high readiness target was to make the most of air force human resources—the young conscripts in training as well as the few civilian technicians employed as full-time professionals. The number of the latter was severely limited by costs and by the shortage of aeronautical technicians in what was then still an agricultural and light-industry country, so the only possible solution was to make the most of the conscript intake. Even before 1948 there had been an aeronautical high school in the city of Haifa, which operated an air force–focused technical school for cadets aged fifteen to nineteen. Dan Tolkowsky, air force commander from 1953 to 1958 (he retired at age thirty-seven) revised the school's program, canceling obsolete airframe skills (including fabric repair for DH Mosquitos and C-47s) to instead focus on electronics, building the skills needed to maintain the avionics of the arriving French fighters, as well as of both homegrown and imported electronic countermeasure and radar equipment.[19] The cadet schools provided the Israeli Air Force (IAF) with personnel already partially trained for demanding technical work in repairing, maintaining, and preparing aircraft for flight. But they also inspired a specific esprit de corps derived from the undoubted reality that the actual combat capability of an air force depends on its operating availability, which depends in turn on the maintainers rising to the task.

That bears directly on the second salient fact of Operation Moked: the second wave of the June 5, 1967, air offensive started at 09:34, less than three hours after the first, a seemingly impossible feat. Once the flight times to the first-wave targets and the return from them are deducted, an average of seven and a half minutes remained from the full stop of the returning aircraft to their stands, to their clearance to roll for takeoff again, rearmed, refueled, and even repaired if necessary.

Using techniques derived from Formula One pit-stop operations, the air force had organized turnaround teams intensively instructed and well tooled for their jobs, but that was only a start. As the Moked plans advanced, fast turnarounds were practiced again and again against a stopwatch, with supervisors and maintenance-team members alike straining to find new sequences, team configurations, or new tooling layouts that would allow them to shave minutes, and later seconds, from the total time of arrival to departure. That is how Israel's small air force seemed so big on June 5, 1967, launching almost 860 individual sorties (475 against airfields, 32 against radars, 119 air intercepts, and 268 air-to-ground attacks) within eleven hours, by achieving extraordinarily high serviceability rates to begin with, and then by ultrafast turnarounds.[20] Over the next five days the IAF conducted approximately 2,790 more sorties—most of them to attack Arab ground forces.[21] When Egypt's president Gamal Abdel Nasser and King Hussein of Jordan jointly concocted the accusation that the June 5 air attacks had actually been mounted by the US and British air forces, triggering anti-American riots all over the Arab world, the sheer number of Israeli aircraft in action gave some credence to the lie.

The next number that needs explaining is the ratio of Arab aircraft destroyed to the number of sorties flown, which was almost one to one, a totally unprecedented success rate for an air attack against ground targets—even without counting the many other targets attacked, including radar stations and runways. The too-obvious answer, that Israel's pilots were exceptionally well trained, was certainly valid—an air force that went so far as to train its ground crews from age fifteen would go to even more unusual lengths to select the best possible pilots and train them as well as possible, for air combat of course, but equally for extreme precision in delivering air-to-ground attacks. Given that the great problem of the air force was the small total tonnage of bombs its fighter-bombers could deliver, the solution had to be to make the most of that scant tonnage by aiming each weapon precisely.

But there was an additional factor that was actually the most important explanation for the inordinate accuracy of the Moked air attacks with guns or bombs: the pilots were instructed to ignore antiaircraft artillery fire in delivering their attacks—and that called for ignoring a great deal of firepower, because Egypt's many antiaircraft units were amply equipped. The reason was historical: reacting to the expected superiority of American airpower, Soviet armed forces allocated disproportionately large resources to their antiaircraft weapons, from a huge number of heavy machine guns both in the common 12.7 mm and the less common 14.5 mm caliber, to the many high-velocity 23 mm cannon in single, twin, and quadruple mounts that accounted for many US aircraft losses in Vietnam along with the slower-firing but longer-range 57 mm guns.[22]

Those guns were concentrated around the Egyptian air bases that were Israel's targets on June 5, 1967. While the bases lacked hardened aircraft shelters and even proper revetments (which can limit sympathetic explosions from one stricken aircraft to the next) they were extremely well defended with abundant antiaircraft firepower, including the first of the Soviet antiaircraft missiles, the S-75 Dvina, better known by the NATO reporting name SAM-2 Guideline.[23] Already well known for shooting down the U-2 flown by Gary Francis Powers over the Soviet Union on May 1, 1960, it saw much action in North Vietnam where it was nicknamed the "Flying Telephone Pole" by US pilots due to its length, the consequence of its two-stage design, with a booster to propel the missile to the high altitudes required for the interception of bomber formations. By 1967 it had brought down 110 US aircraft.[24] But it was only really effective against aircraft at medium or high altitudes, while the plan for June 5, 1967—as almost always with the Israeli Air Force—accepted the higher fuel consumption and terrain collision risks of very low-altitude approaches, except in the case of the most distant targets at Luxor, Ras Banias, and H-3. Moked did not plan any attacks on the twenty-seven Egyptian SAM-2 batteries—the plan was to fly around them or below their effective altitude. Only one Israeli aircraft was shot down by a SAM-2, on the third day of the war. However, once the Arab aircraft threat had been neutralized, twenty-two SAM batteries were added to the target list, with twenty-two strikes to hit them.

Egypt's SAM-2s were ineffective against aircraft coming in at treetop height, but antiaircraft guns were effective. Twenty-six of the forty-six Israeli aircraft lost in the war were shot down by these guns—six while attacking airfields, the rest when conducting ground support—and more were damaged but landed safely. Fifteen Israeli aircraft were shot down by Arab aircraft, fourteen during attacks on airfields.[25] These losses were the direct result of what the pilots had been drilled to do: achieve maximum accuracy by flying straight in and focus only on destroying the assigned targets, with no self-protection by evasive maneuvers. No fighters were diverted to escort others or to fly top cover instead of serving as bombers themselves. There were no onboard electronic defenses, but a few transport aircraft outfitted with electronic equipment did conduct standoff jamming.

That was the real secret of Moked: uncompromised offensive airpower, the very opposite of the US Air Force emphasis on force protection, starting with full-scale SEAD (suppression of enemy air defenses) campaigns before the actual objectives are even attacked. Their aim is to eliminate any remotely possible threat, including obsolete antiaircraft missiles and aircraft, in addition to all potentially operational enemy fighters, missile batteries, antiaircraft guns, and their associated radars and command posts. In the 1991 Desert Storm air campaign, SEAD accounted for some 4,000 sorties, while Defensive Counter

Air, that is, top cover and air defense patrols, came to 5,900 sorties, with another 4,100 flown by allied air forces. All of this was against Saddam Hussein's air force, which was not up to much in air combat even in its best days, and which on that occasion was mostly absent, its best aircraft evacuated to Iran in preference to air combat.[26] The perfectly valid distinction of course is that in 1991 the United States was fighting for Kuwait while in 1967 the Israelis were fighting for their lives, and Moked's high-risk total focus on the offensive was indispensable to deliver a devastating blow with a puny total bombload.

Finally there was the question of intelligence. To deliver bombs and strafe with 30 mm cannon are activities whose actual military value for an air offensive such as Moked critically depended on the quality of the intelligence that guided the entire operation. Even back in 1967 Israeli Intelligence had a high reputation, not only because it extracted valuable intelligence from the otherwise hermetically sealed Soviet Union, in addition to much easier-to-penetrate nearby Arab countries, but also for its secret operations. The most relevant of these for Moked and most recent at the time was Operation Diamond, which delivered an intact Iraqi Air Force MiG-21 F-13 flown by an Assyrian Christian defector to the Hazor Air Base of the Israeli Air Force on August 16, 1966.[27] After extensive flight trials and much simulated air combat, that MiG-21 (tail number 007) was handed over to the US Defense Intelligence Agency, which deemed it important enough to establish a special program, Have Doughnut, to study the aircraft technically, while test pilots of all three services tried their hands with it and against it.[28] (At the time, of course, the United States was still refusing to sell any first-line fighter to Israel, so that precious MiG was something of a propitiatory gift.)

But the Moked air planners were unimpressed by Israeli Intelligence. It kept supplying generalities useful to compile an order of battle with lists of aircraft and air bases, but the strike planners needed altogether more detailed information to allocate as best they could the few sorties they could send against each base, so that pilots arriving at low altitude would immediately know where to find their targets—aircraft on the aprons and in hangars.[29] It did not help that most Egyptian air bases sprawled across vast expanses and were littered with unspecified huts, ex-British bases left over from the Second World War.

Only a veritable cultural change on the intelligence side could satisfy the Moked planners, who needed to know the location of each target aircraft on each base at attack time, just after the routine dawn patrol favored by the Egyptians; the operational readiness of those pilots (not Egyptian pilots in general); their way of life and personal habits; the antiaircraft guns and missiles in place; the thickness and material of the runways, asphalt or cement; the location of the aircraft arming stands, refueling positions, and dummy aircraft; the location of radar stations and the exact scope of their coverage, essential to try to achieve surprise; and more. For the IDF, all this required a revolution in the quantity and quality of intelligence gathering and analysis, starting with the expansion of sources of all kinds—aerial photography and interpretation, communication intercepts, and human scouting—but in the end it all came together only because of one more out-of-the-box remedy: the appointment of fighter pilot Yeshayahu Bareket as head of the Air Intelligence division.[30] Only then was the right perspective acquired—the view through the windshield.

Bareket learned to think as if he were the commander of the Egyptian Air Force. "I want to know what he knows, and at the same time!" he used to tell his men. "I was a fighter pilot and knew very little about intelligence, maybe because I was young and had a lot of chutzpah that I had the courage to ask all the questions and make all the changes."[31] On his first day on the job, Bareket asked his men to inform him how many and which Egyptian planes were flying in Egypt. After a few hours, he received an outdated report. This sparked a revolution that included recruiting new people into air force intelligence and establishing new processes for gathering intelligence. By the end, the IAF's intelligence ability to track Egyptian aircraft movements reached such a level that when four Tu-16 bombers flew from their Cairo air base to distant Luxor on the morning of June 5, 1967, just three hours before the first wave of Israeli aircraft took off toward their targets, IAF intelligence was instantly informed. The pilots were quickly provided with the bombers' new location just in time to change their flight plan to find their targets in Luxor.

Deception was also of the essence for a plan that would have failed catastrophically had the Egyptians been forewarned. First, they were led to believe there were more active IAF airbases than the true number. Second, a routine of launching sorties by Fouga CM.170 Magister jet trainers every morning was kept up, but with the simulated radio signals of first-line jet fighters. The IAF also entered into the realm of electronic warfare to counter Egypt's SAM-2 surface-to-air missiles. The ten biggest air force aircraft, old piston-engined Boeing 377 Stratocruisers, were equipped with frequency-scanning receivers that could trace the SAM-2 batteries' location and operation, to alert the fighter pilots if they were at risk of being "illuminated" (or locked on) almost in real time.

Once the results of the June 5, 1967, air strike were revealed to the world by the photographs of rows and rows of wrecked and burned aircraft, many seemingly destroyed in exactly the same way, enterprising analysts explained that it was all done with infrared missiles, in a clever two-step: the Israelis deliberately warned the Egyptians of their imminent attack, so they would rush to start their engines, thereby creating hot spots for the Israelis' heat-seeking homing missiles. That was just one of least fanciful theories that circulated beyond journalism to enter intelligence assessments. Actually, no precision weapons were employed at all. It was all done with the "Mark One Eyeball" as the saying goes, by pilots who concentrated on their flight paths, ignoring antiaircraft fire as instructed in order to deliver their bombs and gunfire on target, and then further ignored the explosive confusion all around to fly back to their bases and spend as little as seven and half minutes on the ground before flying out to do the same thing again.[32] The only technical innovation had been the homemade runway cratering bombs.

Moked was thus the result of many concerted innovations, tactical, operational, and institutional as well as technical in a small way. It was of course a great success. But ancient wisdom long ago determined that victory is the greatest tragedy, preceded only by defeat, because in victory all that was done appears to be equally fine and equally worthy to be repeated, while only defeat is a discriminating teacher of what works always, what works often, and what works only sometimes, in transiently fortunate circumstances.

The success of Moked aroused Israel's enemies and their Soviet backers to overcome Israeli air superiority. Concluding they could not counter Israeli pilots with improved training or even better aircraft (after 1967 the loss ratio in air-to-air combat increased in the Israelis' favor) they opted for much more passive protection, with ferroconcrete hangars for all aircraft with blast walls between them, and increasingly dense air defenses with the most advanced surface-to-air missiles available, as well as a multitude of antiaircraft cannon. Because the Soviet Union invested much more in developing and producing air defense weapons than the entire Western world, the Egyptians and Syrians could be supplied with the excellent antiaircraft cannon in both static and mobile form (notably the armored, tracked quadruple radar-aimed

23 mm ZSU-23-4 Shilka), as well as increasingly capable antiaircraft missiles that could not be outmaneuvered by clever pilots unprovided with countermeasures.

In October 1973 the Israeli Air Force's attempt to aid the ground forces responding to the Egyptian and Syrian surprise offensives collided with air defenses it could neither destroy nor avoid, resulting in unsustainable losses. While the unsupported ground forces coped as best they could, the airmen faced defeat, until rescued by the advancing ground forces that overran Egyptian and Syrian air defense batteries. That reversal of fortunes had a high cost in casualties, and did not diminish the bitter sense of defeat of the airmen, which would power their revenge.

Artzav 19, the 1982 Surprise

"In 1973, we [the air force] failed the people of Israel, we had to recover their trust," said Brigadier General Aviem Sella.[33] On June 6, 1982, following escalating attacks on Israel by Palestinian military units on Lebanese soil, IDF ground forces invaded southern Lebanon in great strength. The objective was to drive out both the Palestinian and Syrian forces occupying the country. (Israeli troops were welcomed as liberators by all communities, but Shi'a attitudes would change when they lingered.)

On June 9, 1982, from 14.00 to 16.00, Israel's Air Force launched operation Artzav 19 ("Mole Cricket 19"), whose ambitious aim was to destroy in a single attack all Syrian surface-to-air missile batteries in Lebanon, which were fully integrated in a state-of-the-art Soviet-supplied air defense system, along with the most effective antiaircraft guns ever made and a large force of Syrian Air Force jet fighters. What ensued was the world's first combined attack by manned and unmanned aircraft against an integrated air defense network equipped with missiles and guns, which, when the Syrian Air Force intervened in great numbers, led to the largest single air battle since the Second World War.

The initial Israeli attack destroyed nineteen antiaircraft missile batteries—including some SA-8s, the most advanced Soviet missile of that class—and also destroyed twenty-six Syrian aircraft. Over the next two days, five more batteries were destroyed, and the number of Syrian aircraft destroyed in air combat reached a total of eighty-two, with five more shot down by fire from the ground.[34] An additional six SAM-6 batteries were also destroyed.[35] As the

Syrians brought up reinforcements, a further eight SAM-6 batteries were destroyed during the month of fighting that ensued. There was not a single Israeli air loss, in what amounted to one of the most one-sided battles in history.

At the time, Israel's air victory attracted much attention because the Soviet antiaircraft missiles that had been so formidable just nine years earlier in 1973 seemed to have suddenly become ineffectual, even though some were in fact far more advanced than their predecessors. But that actually distracted attention from the real feat accomplished on June 9, 1982. Instead of a maximum-risk, all-out attack, as in Moked in 1967 with almost every pilot and every fighter committed, Artzav 19 was a rather small attack: the Israelis flew a mere 125 strike sorties in all, with 56 support sorties, a low number indeed, which nevertheless sufficed to destroy the densest air defense network in the world, with the possible exception of the Moscow regional air defenses.[36] The annihilation of the up-to-date Soviet defense system was a momentous event that sent shock waves through both the Soviet and the American defense establishments. In Moscow there was deep distress, with some already viewing the defeat as the beginning of the end of the Soviet empire, because it seemed that the Soviet Union could never catch up technologically in the new information age.[37]

Just like the Moked surprise attack, Operation Artzav 19 was meticulously planned and rehearsed. But while Moked stretched the IAF's inventory to its very limits, Artzav 19 was so efficient that it was completed in just two intense hours; fighter-bombers already up in the air and circling in readiness for the second-wave attack had to drop their weaponloads at sea.[38] Clearly this was not a victory gained just by sheer talent and intense commitment but rather by something conceptual and something technological that was entirely new. But that conclusion was obscured by misleading early reports that were never officially corrected; the operational details remain official secrets to this day.

The nine-year interval from 1973 to 1982 would suggest there was plenty of time for this instance of innovation, but in fact the Israeli Air Force had much less time than that because the development and delivery of the necessary equipment from the United States stretched over several years, and even then did not include the most up-to-date equipment the IAF wanted. The IAF did not want some of the most advanced US equipment offered, including gliding bombs (which did not offer the option of manual in-flight guidance) and the dedicated Wild Weasel airplanes equipped exclusively for air defense suppression that contradicted Israel's need for multirole fighters, such as the Mirage IIIC had been before the arrival of the F-4E Phantom, F-16, and F-15.[39] As result, the Israelis were forced to design and manufacture much of their own air defense suppression materiel, starting from scratch in most cases. Because the overall method was to employ a variety of overlapping technological approaches, some never tried before, rather than rely on any specific system, the procurement challenge was central to the overall effort, doctrine-guided though it was.[40]

This sequence of events started in 1963, when the first Soviet antiaircraft missile the Israelis ever encountered—the S75, designated SAM-2 by NATO—was first deployed in Egypt. As noted earlier, its operational performance during the June 1967 war was unimpressive because the Israelis had identified the location of all twenty-seven Egyptian SAM-2 batteries and could route their aircraft around them, or fly nap-of-the-earth profiles below the minimum engagement altitude of the SAM-2.[41] Experimental standoff electronic warfare systems were employed but their value was uncertain.[42] Only one Israeli aircraft was destroyed by a SAM-2.

Right after the 1967 war, which was a major defeat for Soviet weapons and doctrines, tens of thousands of Soviet advisors were sent to Egypt, Syria, and Iraq to rebuild their forces, with 20,000 advisors operating in Egypt alone, including many air defense specialists.[43] The rebuilding of Egypt's military capabilities coincided with increasing but intermittent skirmishing that rose and fell in intensity from July 1, 1967, until September 1969. Fighting included small-arms fire exchanges (the rival forces were barely 150 meters apart on either side of the Suez Canal), artillery and tank duels, and reciprocal cross-canal ground raids and ambushes as well as reciprocal air strikes and aerial duels. It was during this War of Attrition that the Israeli destroyer *Eilat* was sunk in October 1967—the first warship ever sunk by a naval missile.

The IAF's superiority in air-to-air combat was quickly proven—in three years of air-to-air duels, July 1967 to August 1970, it lost only six fighters in air-to-air combat while shooting down 113 Arab fighters and bombers, 86 of them Egyptian. The Soviet and Egyptian response was to multiply ground-to-air capabilities, both missiles and guns. However, until March 1969 no Israeli aircraft were destroyed by surface-to-air missiles (SAMs), which rarely forced Israeli aircraft to abort strikes to evade them. In spite of their growing number Egypt's SAM batteries could not provide continuous coverage of the front, so the IAF struck targets outside SAM cover or, if the designated targets were within SAM cover, standoff jammers and complicated air maneuvers by aircraft flying on different vectors at different altitudes were enough to surprise and confuse the Egyptian SAM crews. When SAM launches did occur, they were usually detected, allowing the aircraft they threatened to conduct aggressive aerobatics to evade the missile. Antiaircraft guns proved more deadly to aircraft flying close to the ground to evade radar detection, prompting a preference to approach targets higher up and thus within the SAM-2 envelope but above gun range.

From March 8, 1969, the Egyptian attacks intensified drastically—thousands of shells were fired on Israeli positions every day, tens of thousands on some days. The next day, a small observation aircraft became the first Israeli aircraft loss to the SAM-2 since June 1967.[44] By late June it became clear to the Israelis that they could not sustain the completely lopsided artillery duel with the Egyptians (a few dozen Israeli guns against more than 1,000 Egyptian). It was therefore decided to drastically increase the IAF's participation; from July 1969 until the cease-fire on August 7, 1970, the IAF conducted 8,200 strike sorties that dropped approximately 50,000 munitions on 683 air defense targets, 1,353 ground force targets, 180 military infrastructure targets, and five naval vessels.[45]

The decision to escalate the tempo of strikes required more direct action against Egypt's air defenses, which were vigorously reinforced to engage Israeli aircraft. The Israeli response was no longer to avoid and evade, but to directly attack and destroy SAM-2 batteries. On July 20, 1969, a SAM-2 battery west of Port Said was destroyed; two days later, in a larger operation, attacks were aimed at SAM-2 batteries in Abu-Suweir, Ganifa, Al-Menif, and Al-K'hafir.[46] Most Israeli anti-SAM strikes were executed by small batches of aircraft combining high and/or low approaches with the standoff jamming of Egyptian radars and communications. But a few were massive strikes, with dozens of aircraft simultaneously attacking multiple SAM-2 batteries along with the antiaircraft gun batteries deployed around them.

For eleven months, the duel between Israeli airmen and Egyptian air defenses was unceasing. The Israelis were destroying SAM batteries, antiaircraft guns, and radars to preserve air supremacy above the Suez Canal, which was essential to deter Egyptian crossings and enable strikes on Egyptian artillery and infantry harassing Israeli ground forces. The Egyptians were striving to push forward more missile and gun batteries closer and closer to the Suez

Canal to limit the Israelis' freedom of action. Despite the doubling of Egyptian air defenses to fifty SAM-2 batteries and 1,000 guns, from July to December 1969 no Israeli aircraft were lost to SAMs, though a few were lost to antiaircraft guns. On December 24, 1969, a Mirage on a photographic reconnaissance mission was surprised by a missile coming through thick clouds, but the pilot landed safely and got away before the damaged aircraft exploded on the ground.[47]

Israel's campaign against Egypt's air defenses was not confined to electronic attacks and air strikes: in December 1969, a commando raid seized a then-advanced Soviet P-12 Yenisei (NATO's Spoon Rest A) radar complex.[48] The aim was to study its electronic characteristics in order to improve countermeasures. To retrieve the heavy radar components the raiders relied on powerful CH-53 helicopters newly supplied by the United States, whose immediate reward was access to the P-12, then Vietnam's best search radar deployed against US aircraft.

Meanwhile, in September 1969 the first American-made Phantoms had arrived. Unlike the Skyhawks and French aircraft, they had integral radar-warning systems. After the destruction of the Mirage, the air force decided that only Phantoms would conduct strikes against the SAMs as they did not need to see the missile's launch to prepare to evade it. An external electronic-countermeasure (ECM) pod was developed, to be carried by Vautours and Skyhawks; the Mirages and Mystères could not carry it. When they were needed to patrol SAM-threatened areas they were accompanied by a Phantom to provide warning. The number of standoff ECMs was increased and they were used in every mission. However, experience showed that even if the jamming effects might reduce the threat, many missiles still got through and had to be evaded by sharp aerobatics.

As IAF strikes inflicted heavy casualties on Egyptians in manpower, artillery weapons, antiaircraft forces, and aircraft, Nasser appealed to Moscow for help. The Soviets responded by sending their own best air defense pilots to participate. Preparations for this move began in August 1969, the final decision to intervene was made in December 1969, and the first Soviet units arrived in March 1970.[49]

Meanwhile, Israel tried to increase the pressure on Egypt, seeing an opportunity in the regime change from Nasser, who died in September 1970, to Anwar Sadat. Operation Prikha ("Blossom") attacked military bases and SAM sites deep in Egypt, even around Cairo itself, to show the population at large that Sadat's government could not even defend the capital. Although all the targets were military and not strategically very significant, the bombing caused strong public reactions, including episodes of mass panic—especially after two civilian buildings, mistakenly identified by Israeli pilots as a weapons factory and a military headquarters, were heavily bombed.[50]

From January 7 to April 13, 1970, Operation Prikha included eighty-eight sorties, and no Israeli aircraft were shot down. However, it did not achieve its strategic goal of pressuring Egypt's leaders to deescalate the fighting across the Suez Canal. Worse yet, at the time, the Prikha strategy was thought to have aggravated Israel's situation by rousing the Soviets to intervene. This assessment was proven wrong only after the dissolution of the Soviet Union, when declassified Soviet documents showed that Moscow's decision to intervene had preceded Prikha.

The effort allocated to Prikha did not diminish Israeli air strikes on Egyptian frontline forces, whose air defenses continued to fail in halting the continuous Israeli strikes. But in March 1970, a large Soviet airlift delivered the troops and headquarters of the Eighteenth Special Antiaircraft Missile Division, with no fewer than seventy-two batteries of SAM-3 surface-to-air missiles, complemented by 23 mm antiaircraft guns and shoulder-launched SAM-7 Grail missiles for use against low-level intruders. The Soviet 135th Fighter Aviation Regiment also arrived, with ninety-five of the most advanced versions of the MiG-21—the MiG-21MF interceptor—and fifty Sukhoi-9 interceptors. Along with their command posts, headquarters, radars, and electronic warfare units, these forces formed a complete air defense system. Initially they were deployed to defend only Cairo, but gradually they began to push eastward toward the frontline on the Suez Canal. Not willing to clash directly with the Soviet Union, the Israeli government ordered a cessation of all strikes into deep Egyptian territory.

The Egyptians provided strongly fortified infrastructures for the Soviet forces by means of an accelerated, large-scale construction effort that was itself defended from attack by overlapping surface-to-air missile batteries. Even before they understood the full scope of the Soviet intervention and the purpose of the vast construction effort they were seeing, the Israelis immediately started bombing the construction sites. But the Egyptians kept rebuilding, despite thousands of casualties.

The SAM-3s and the armored, tracked quadruple ZSU-23 gun vehicles could move quickly from site to site to set ambushes. For that purpose, the

Egyptians built three air defense sites for each battery, for both deception and survivability.[51] Some empty sites were fitted with wooden dummies of SAM systems, including pyrotechnics and electronic signaling, to make them appear lived in.[52]

Moreover, the Soviet divisional-level command-and-control net allowed several batteries to be subjected to a single battery's radar, thus enabling the launching of missiles from several batteries whose own radars were silent in order to surprise incoming Israeli aircraft, a tactic enhanced by the easy mobility of the SAMs between different sites. The dense battery deployment with its overlapping fields of fire formed an array so dense that Israeli pilots spoke of a "missile wall."[53] Yet the air force continued to attack the SAM sites and their radars to enable strikes on Egyptian ground forces. Then on April 12, 1970, Israeli F-4 Phantoms deliberately attacked SAM-3 batteries known to be manned by Soviet troops; the batteries were destroyed before they had a chance to launch any missiles at attacking aircraft.

In June 1970 all Egyptian air defense units were subordinated to the Soviet Air Defense (PVO Strany) commanders in Egypt, who concentrated and redeployed the combined air defense forces to provide continuous coverage with overlapping missile batteries all the way from Cairo to within sixty kilometers of the Suez Canal front line. More aggressive than the Egyptians, the Soviet operators would advance the SAMs in swift actions instead of incrementally.[54] On the morning of June 30, 1970, the complete Soviet missile array became active and started launching SAMs at Israeli airplanes.[55] On that afternoon, when the IAF counterattacked the most forward SAM sites, the pilots discovered that the rules of the game had changed: not one or two or several but rather dozens of missiles were launched against each formation.[56] Two Israeli fighters were lost to missiles, shocking the airmen. The immediate response was to try new evasive techniques, team tactics, and electronic countermeasures to reduce the effectiveness of the Soviet missiles, preferably without flying very low and therefore within range of the antiaircraft guns that had caused the majority of Israeli air losses in the past.

The new tactics, first implemented on July 5, 1970, were to conduct massive, closely choreographed, four-wave, sixty-aircraft attacks against the forward SAM batteries instead of the previous separate attacks by smaller formations. The new tactic was highly dependent on precise timing in order to saturate the defenses with multiple aircraft approaching from various directions and heights with various flight patterns. Results were mixed, and a third aircraft was lost to the SAMs. At this point the most forward SAM-2 batteries were only thirty-five kilometers from the canal front line, with the SAM-3 batteries some forty-five kilometers from the canal. At that point, the Israelis reverted to small strikes, combining electronic jamming with "seduction" flights to draw enemy fire and sudden strikes by aircraft waiting beyond radar coverage to attack batteries that had just launched missiles and therefore could not be ready to launch again.

On July 18, 1970, the IAF again attacked the Soviet missile array in a large-scale operation, Etgar ("Challenge"). It was the first time the Israelis used the US-supplied AN / ALQ-71 electronic-countermeasure jamming pod (code-named Afunah Reikhanit, "Scented Pea").[57] The US experts on the system advised IAF pilots to fly straight into the missile zone in a steady formation without performing evasive maneuvers so as to maintain effective mutual coverage (in a "pod formation").[58] But both the technology and the tactic failed, as three precious F-4 Phantoms were hit, two of them were destroyed, and a revered squadron commander, Shmuel Khetz, was killed.

It turned out that the pod was partially effective against SAM-2s, but totally ineffective against SAM-3s. The Israelis continued to use it, but reverted to evasive aerobatics instead of depending on the gadget alone.[59] Initially, despite the cost, the strike results seemed satisfactory, with seven SAM batteries apparently destroyed, but later it was discovered only three were active batteries while the rest were decoys.[60]

With the Israelis struggling to destroy the SAMs, the initiative passed to the Soviet air defense commanders, who started scrambling Soviet-manned fighters to intercept Israeli aircraft. On July 25, 1970, Soviet MiG-21s intercepted Israeli A-4 Skyhawks on a ground attack mission and chased them back into Israeli-controlled Sinai airspace. As subsonic aircraft the A-4s were outclassed by the MiG-21s, and one A-4 hit by an Atoll air-to-air missile was forced to land at the forward air base in Rephidim. The Israelis responded in kind on July 30, 1970, with Operation Rimon 20, an air ambush in which twelve Mirage IIICs and four F-4E Phantom IIs lured and caught a Soviet reaction force of twenty-four MiG-21MFs. In the ensuing air battle, which was more confused than most, the Israelis suffered one damaged aircraft that landed safely, while five MiG-21s were destroyed and four Soviet airmen killed. As with strikes on Soviet ground personnel, the Israelis reported the destruction of Egyptian rather than Soviet aircraft, so that they could not be contradicted by Soviet spokesmen, who kept denying any active involvement in the fighting. Nor did Moscow react at the diplomatic level, or even denounce the attack. Instead, there was a further advance of SAM-3 batteries toward the Suez Canal.

But the message was received in Washington as well as in Moscow, and the result was a US-mediated cease-fire agreement signed by Israel, Egypt, and the United States (the USSR did not sign it), which came into effect on August 7, 1970.[61] The agreement was violated almost immediately by the Egyptians and Soviets, who further advanced their missile batteries toward the canal zone. The United States did not try to renounce or to enforce the cease-fire, reacting instead by supplying Israel with all the weapons it had developed to fight air defenses: radar-warning receivers to be fitted to aircraft, chaff (to confuse radars), flares (to confuse infrared sensors) with their respective dispensers, and AGM-45 Shrike missiles that homed in on radar emissions as well as CBU-24 cluster bombs that increased the probability of hitting targets. All this equipment was a consolation prize for Israel in exchange for its acceptance of the cease-fire violations instead of renewing war, even as Egypt was openly breaking the agreement by moving forward the SAM batteries, whose missiles could soon threaten even aircraft flying well within Israel's side of the Suez Canal.

Between the August 1970 cease-fire and the October 1973 war there were a number of missile launches aimed at Israeli aircraft flying in Israeli airspace. In one case a Stratocruiser carrying signal intelligence equipment and flying a route deemed safe was shot down by a SAM-2 battery that had been secretly advanced to a new location. A retaliation strike with Shrikes failed. This anticipated what would happen at the start of the October 1973 war, when Egyptian SAM batteries would inflict heavy casualties, neutralizing the effectiveness of Israel's airpower.

During those three years, the IAF was trying to prepare for the next round because after the US refusal to punish, or even acknowledge, the cease-fire violations, nobody believed that another war could be avoided. From July 1967 to August 1970 a total of sixteen Israeli fighters were destroyed on the Egyptian front, six hit by surface-to-air missiles, five of them within five weeks. Another six Israeli aircraft had been shot down by Egyptian antiaircraft guns and four by Egyptian fighters. Eighty-six Egyptian combat aircraft were shot down by Israeli fighter aircraft during the same period. A further fifteen Israeli aircraft were lost on other fronts.

The major lesson learned was a reversal of priorities—the first act of the 1967 war had been to eliminate Arab combat aircraft (especially the bombers that threatened Israel's civilian rear) while ignoring Arab antiaircraft defenses. But the first act of the next war would have to be the destruction of the SAMs, which would require a mix of technology, techniques, and tactics yet to be developed. The effort started right away, yet its results would not arrive in time for the October 1973 war.

The anti-SAM campaign plans, one for the Egyptian front and one for the Syrian front, were inherently much more complicated than Operation Moked had been in 1967. The basic concept was the same for both plans: Dugman ("Fashion Model") for Syria and Tagar ("Challenge") for Egypt. Each required exactly timed sequences of standoff electronic countermeasures (jamming and chaff clouds), drone decoys to draw fire, long-range artillery fires to destroy or at least interfere with the most forward SAM batteries, salvos of Shrike antiradiation missiles to hit battery radars, and then a series of low-level flights to attack antiaircraft gun positions to open safe low-altitude routes for the aircraft sent to attack the key targets, the SAM batteries.

Given that the various electronic-countermeasure systems and the Shrike missiles had all been tried but had not proven themselves successful, and that new SAMs, especially the SA-6s, could function in spite of the available ECMs, the above measures were considered useful but not sufficiently effective to serve as the main protection of the attacking aircraft, which would have to rely on superior flight tactics. For this purpose, each attack wave of strikes would combine different flight methods. Initial strikes would employ the safer but less accurate Kela ("Slingshot") technique—in which aircraft would fly low and fast until, at a predetermined point before entering antiaircraft gun range, the pilots would raise the nose of the aircraft to a prescribed angle and release the bombs, then turn to egress a very low altitude; the bombs would fly in a parabolic pattern to their targets. The pilots could not see the targets they were aiming at, but if the computations and the pilot's hand were accurate, enough targets would be hit to enable the riskier but more accurate technique of Hataf ("Hurried"): aircraft would fly below the radar threshold to a predetermined point near the target, pop up a few thousand feet, roll over, see the target, aim and dive on it, release bombs, and egress at very low altitude. To evade missile fire, the entire pop-up-to-egress sequence was to last only seconds, less time than it took the battery to engage the aircraft.

This tactic, however, had one critical weakness: the pilots could not see their targets until they popped up and rolled over, and would have no time to visually search for the target if it was not where they expected it to be. Therefore, for both Kela and Hataf techniques, the air strike planners had to know in advance precisely where each enemy gun or missile battery would be. This in turn required a photographic reconnaissance flight prior to attack, with the hope that the batteries would not be moved between the reconnaissance flight and the strike, even though to move with some frequency was a standard countermeasure routinely practiced by the enemy.

But the crippling shortcoming of contemporary photographic reconnaissance methods for that specific purpose was their slow pace. It was eight hours from the moment the photograph was taken to the moment bombs were to be released on target, adding up the flight time to return to base; the time needed to transfer the film from the aircraft to the developers and from them to the analysts; the time needed to analyze the imagery; the time needed to transfer the results to the planners; time for the planners to digest the data, create a plan, and send it to the participating squadrons; time for them to study it and assign pilots and aircraft; and time for each pilot to study his mission, board his aircraft, and fly to the target. For the bulky, difficult-to-move SAM-2s an eight-hour lapse was not too much in most cases; for the more easily moved SAM-3s it would mean that some batteries would have been moved, and the SAM-6s could be moved twice within eight hours. Therefore IAF planners preferred a surprise preemptive strike, as in June 1967.[62]

One organizational weakness was that photographic analysis to determine the exact location of the missile batteries, radars, and launchers was not performed at air force headquarters, but at Military Intelligence headquarters, where all aerial photography was analyzed. The consequence was additional delay—extra time to transfer the results from the imagery analysts to the air force planners. In a few years all those functions would be performed in near real time, regardless of location, but not in 1972–1973.

Israel's enemies did not stand still. The apparent success of the air defenses in the summer of 1970 became the basis for their planning, and they invested greatly in ground-to-air forces, continuing even when the Soviet combat forces departed, leaving only advisers and instructors. By October 1973 the Egyptians had 146 missile batteries (72 high-altitude SAM-2s, 64 medium-altitude SAM-3s, and 10 low-to-medium-altitude SAM-6s), of which 55 (25

SAM-2s, 20 SAM-3s, and 10 SAM-6s) were deployed near the Suez Canal, and could therefore operate over some distance into Israeli airspace, while the others further back protected Cairo and rear areas. With a smaller front to defend, the Syrians could achieve the same density with fewer batteries: 36 (13 SAM-2s, 8 SAM-3s, and 15 SAM-6s), of which 25 were deployed at the front (7 SAM-2s, 3 SAM-3s, and 15 SAM-6s). Nine batteries around Damascus were close enough to the front to help protect the rear units of the Syrian ground forces, while two more protected Dmeyr airfield farther back.

Both the Egyptian and Syrian armies also deployed some two thousand antiaircraft guns of all types and hundreds of man-portable SAM-7 launchers. As for the SAM batteries of both armies, they each had multiple fortified sites (following specific Soviet directives), some with dummy equipment when the battery was elsewhere—good enough to have fooled the Israelis. Moreover, there were radar emitters around each site to seduce Shrike missiles, and batteries of antiaircraft guns and also man-ported SAM-7 teams to protect against low-flying aircraft. Battery locations were selected to form overlapping zones, and integrated central command systems allowed the multibattery engagement of each target.

Before the outbreak of war on October 6, 1973, Minister of Defense Moshe Dayan and IDF Chief of Staff David Elazar had agreed that in the event of war, the IDF should focus on defeating the Syrians before diverting the bulk of their forces to the Egyptian front, because of the short distance between the location of Syrian ground forces and the Israeli villages and towns on the Golan Heights and in the Hulah Valley just below it. By contrast, on the Sinai front some 150 kilometers of empty desert separated the Suez Canal front line from the nearest Israeli civilian settlement. Another reason to prioritize the Syrian front was that Dayan and Elazar hoped a quick victory over the Syrians would deter Jordan and Iraq from joining the Syrian-Egyptian offensive; they further decided that in the event of war the Operation Dugman suppression campaign against the Syrian SAMs would begin immediately.

As the next war approached the IAF was confident it had the solution. However, when it burst upon them, the opening conditions of the October 1973 war were completely different from those envisaged. When the definitive warning came on the morning of October 6, IAF commander Benny Peled requested permission to launch a preemptive strike and was refused. Needing American backing, Israel could not allow itself to be portrayed again as the aggressor. He had been confident permission would be given, and the IAF had been preparing its aircraft since early morning. The refusal required a complete reappraisal of the situation and a complete change in aircraft loads.

However, even if permission had been given, a second critical vulnerability manifested itself: the skies over the expected battlefield were too cloudy. Targets could not be seen by the pilots. Instead of launching Dugman, Peled decided to attack the Syrian air bases beyond the clouded area. That decision meant that all the aircraft had to unload their anti-SAM munition loads and replace them with a mix suitable for striking air bases—concrete-penetrators to crater runways and attack ferroconcrete aircraft shelters—and the pilots had to study new missions. Given the expected Arab H-hour, 18:00 GMT, there seemed to be sufficient time to conduct the transition.[63]

Then came a second surprise—at 13:55 the ground forces began reporting massive artillery bombardments and IAF radar operators saw a mass of Egyptian and Syrian aircraft approaching the borders. The IAF was caught in midtransition from one ground-attack mode to another, neither compatible with the interception of incoming intruders. Reacting to the terrible danger of a reverse Moked that might cripple the air force, Peled ordered all aircraft into the air. No attacks on Israel's major air bases ensued, but next came a scramble to bring back the aircraft and prepare them to respond to the urgent ground-force requests for air support. Without the reserve units just being mobilized, the forces at the front were facing enormous odds. They needed air support immediately; they could not wait for the IAF to first achieve air superiority.

By the late evening of October 6, Israeli fighter-bombers had conducted some 200 air-to-air sorties, shooting down sixteen Egyptian helicopters landing commandos behind Israeli lines and some twenty Egyptian and Syrian fighter aircraft. Another 110 sorties struck Arab ground forces, but they had to fly into heavy antiaircraft fire. Six aircraft were lost and others left damaged but repairable. Dozens of sorties were also launched that night. Meanwhile, the IAF prepared to launch Dugman the next morning. Initially the Israeli forces on the Syrian front seemed to be holding out, whereas those in Sinai were overwhelmed, so the IAF was suddenly ordered to launch Tagar instead of Dugman.

On the morning of October 7, the first of the three planned waves of Tagar was duly launched and seemed to be successful. But then the IDF's overall situation reversed; during the night Syrian forces had penetrated between the sparsely deployed Israelis and were advancing rapidly across the Golan toward the Israeli civilian population in the Hula Valley. Most of the reserves were still en route to the fronts, and in the Golan the thin forces in place were suffering heavy casualties. By contrast, in Sinai the situation seemed better.

The air force was therefore ordered to provide immediate support on the Golan.[64] Tagar, aimed at Egypt's SAMS, was halted in midstride so that Dugman could be hastily prepared, while fifty-five sorties were launched into the teeth of Syria's undiminished defenses to attack the advancing armored and mechanized forces. Most of these sorties conducted Kela strikes for survivability, but caused very little damage, while some attempted more precise strikes that met heavy antiaircraft fire.[65] The exact results of these sorties could not be determined, and it is uncertain if they were effective at all in slowing down the Syrian advance to gain time for the reserve formations just beginning to arrive to block the routes from the Golan Heights to the Hula Valley, where many civilians were in imminent danger.

The decision to conduct Dugman without any advantage of surprise or proper preparations was made in that dramatic context: if the Syrian SAMs could be neutralized, the IAF would be able to conduct mass precision bombing attacks on the Syrian ground forces and change the balance on the ground. It was a high-stakes gamble in desperate circumstances. But Dugman failed completely. There had not been enough time to conduct a photographic reconnaissance to update battery locations, which had all moved since the last update the previous day. Only one SAM battery was destroyed because a pilot happened by chance to see the new location as he rolled over to dive on his given target, which now, like all the other targets, was just empty revetments. Another Syrian SAM battery was partially hit and returned to action two days later. Furthermore, the original egress routes had been planned on the assumption that the Syrians had not yet penetrated the Golan. But they had, taking large numbers of antiaircraft guns with them whose locations were entirely unknown; the aircraft returning from Dugman flew low over them and paid the price.

Tagar had been a partial success, cut off in midstride, but since it was not completed it is difficult to assess whether it would have achieved the required result. It too had not been conducted exactly as planned. In retrospect it is clear that Dugman and Tagar had failed because of their complexity and resulting lack of flexibility in the face of adverse conditions: both political—the disallowed preemptive strike—and military, the urgency of air support for the grossly outnumbered ground forces, with the October clouds being another obstacle. Too many things had to converge for the two plans to work—an inherent defect for any military plan. Moreover, in addition to adverse circumstances and the inherent lack of flexibility of each battle plan, there were also both operational and even conceptual errors. It was Moked in reverse, with everything going wrong instead of the other way around.

For example, having decided to begin the war with Dugman, the IAF sent its one decoy-drone unit to the Golan. It was equipped with US-made BQM-74 Chukar target drones repurposed to serve as decoys under the code name Telem. But when Dugman was canceled on the night of October 6 and the decision made to execute Tagar against Egypt's SAMs at dawn on October 7, the decoy-drone unit was up on the Golan. Furthermore, in the confusion, the Telem unit commander was not informed of the change of plans, so at the allotted time, as per his original orders, he launched a salvo of Telems toward Syria. He and his men witnessed no fewer than twenty SAM launches against their four drones, many launched from batteries whose presence and location were unknown until then. The unit commander immediately called air force headquarters to report that the Telem decoy drones were functioning very well—they had never been tested before in order to preserve the element of surprise. But he was then informed that unfortunately the IAF was not executing Dugman on his front but Tagar in the south, well beyond his range.[66] When Dugman was finally conducted a few hours later, the unit had no drones left.

Another failure came with the planned artillery support for both takedown plans. In Sinai there were not enough long-range guns to conduct the planned bombardment; those in place were busy responding to emergency fire-support requests from beleaguered ground units—Tagar had to do without. On the Golan, some of the artillery deployed for the original plan had been forced to withdraw and the remaining artillery units were heavily engaged in support of the ground forces. Although they did manage to fire some salvos, as had been planned for the previous day, the confusion created by the nighttime switch to Tagar caused them to fire too early. In fact, it was perhaps this fire that prompted the Syrians to relocate their batteries so that a few hours later, when Dugman was indeed implemented, the pilots bombed empty positions.[67] As for the electronic-warfare units, they had deployed to the Egyptian front and participated in Tagar, but were unable to transit in time to the Syrian front to participate in Dugman. While still trying to assess the results of the anti-SAM attacks in Syria, and before discovering their utter failure, the air force executed another 40 attack sorties in the Golan and 140 more in Sinai in an attempt to destroy the pontoon bridges on which Egyptian forces were crossing the Suez Canal.

Altogether, on October 7, the IAF lost twelve aircraft on the Syrian front: six in the failed anti-SAM strike, five while attempting to provide ground support, and one to Syrian antiaircraft artillery while chasing a low-flying Syrian aircraft. A further ten aircraft were shot down over Sinai and Egypt, for a total of twenty-two Israeli aircraft lost with nothing gained. This would be the worst day of the entire war, but the air force chiefs did not yet know that. What everyone in the IAF did know was that such a loss rate could not be sustained.

Even after the failure of Dugman, new air force photo reconnaissance sorties over the Golan again failed to identify and locate ten of the fifteen SAM-6 batteries the Syrians were known to have. It was only after extensive research conducted two years after the war that air force analysts were finally able to identify thirteen of the fifteen SAM-6 batteries in the photo strips dating back to October 5 and 7, 1973.[68] In other words, the overall system was just too slow to keep up with mobile targets.

The first decision implemented immediately by the IAF was to forgo large-scale operations to destroy the SAM threat in order to provide clear skies. Instead it turned to much smaller operations to nibble at the SAMs, while diverting its focus to supporting the ground forces despite the SAM threat. In total during the entire war the IAF conducted some 1,400 sorties against SAMs (approximately 220 in Tagar and Dugman), during which only three Syrian SAM batteries were destroyed and five damaged against thirty-two Egyptian SAM batteries destroyed and eleven damaged. A further eleven Egyptian SAM batteries were destroyed by Israeli ground forces after the IDF crossed the Suez Canal to raid on the Egyptian side with tanks and mechanized infantry. But at no time, until the last two days of the war, did the IAF achieve complete operational freedom of action to bomb enemy forces at will—every single strike required a struggle to penetrate air defenses, thereby reducing the overall combat value of Israel's airpower.[69]

In spite of that, the majority of IAF sorties during the war were executed to provide support to the ground forces. For example, on October 11, as the IDF Northern Command counterattacked Syria, it received 221 supporting ground-attack sorties with another 130 on the following day. Concurrently,

41 sorties on October 11 and 11 sorties on October 12 attacked Syrian SAM batteries. Those missions, with others in Syria against air bases and major infrastructures, and air-to-air combat to prevent the Syrian Air Force from attacking Israeli ground forces, cost eight aircraft on October 11 and three more on October 12.[70] All told, the IAF conducted approximately 5,270 ground-support missions—most of them interdiction sorties with the rest being close-support.[71] But because the SAM threat was never completely neutralized, the effectiveness of that air support was reduced and the cost in lost aircraft was high. Another 3,180 sorties were conducted to patrol or to escort strike aircraft, shooting down approximately 260 enemy combat aircraft and 35 helicopters. Israel's Hawk missiles and antiaircraft guns also shot down some 50 combat aircraft and 15 helicopters.

The war cost the IAF a total of 102 combat aircraft, 5 helicopters, and 2 light aircraft, with roughly one aircraft lost per 110 combat sorties.[72] Of the combat aircraft lost 57 were shot down during the first five days of the war, the remaining 45 lost over the next 14 days.[73] Approximately half the combat aircraft lost were hit by SAMs—a ratio of approximately forty missiles launched per aircraft destroyed.

[11]

Airpower Restored with a Technological Leap

FOLLOWING THE 1973 DEBACLE, EZER WEIZMAN FAMOUSLY SAID, "The missile has bent the aircraft's wing."[1] The trauma of the 1973 war had challenged the very ethos of the victorious air force of 1948, 1956, and 1967. "We felt humiliated," said fighter ace Aviem Sella, who played a key role in leading reforms after 1973, adding that "we were determined to find a way to restore our professional honor . . . the Yom Kippur [war] exposed a number of gaps that the air force had to fill if it wanted a better outcome in the next round."[2]

Some officers believed that the air force had been outclassed technologically by Soviet antiaircraft missiles. Others did not agree that technology was the core issue. Moshe Dayan notably insisted that the solutions to the problem would not be technological ("electronics will not win the war") but tactical and operational: it was a question of fighting in a clever and daring way, relying on the intelligence of the warriors themselves.[3] But it was undeniable that the Soviet-supplied missile umbrellas had very severely restricted air operations throughout the war and also that the available electronic countermeasures and standoff weapons such as Shrike and the gliding bombs could not neutralize them, thereby compelling the Israeli Air Force (IAF) to send its aircraft deep into the maw of the enemy air defenses, losing a quarter of them to surface-to-air missiles (SAMs) and antiaircraft guns.

After the war, the air force conducted a comprehensive investigation, identifying a long list of major deficiencies—and possible solutions for the future. In retrospect it was the planning that had been off course: Dugman and Tagar were both rigid and fragile plans because they were entirely dependent on exact knowledge of the locations of each missile launcher. When the launchers' mobility made that impossible given the slowness of the intelligence process at the time, the planning process left no room for any alternative tactics.

In response to these findings, the air force concluded it was essential to shorten the interval between the detection of SAM batteries, their identification as real operating batteries rather than simulations, and their effective attack, before they moved again.[4] Also, it was necessary to find ways of reducing the risk to aircraft conducting the strike, whether from the SAMs themselves or from their protective array of antiaircraft guns. To achieve dramatic improvement on both fronts required a variety of innovations in technology, work processes, and tactics. The central goal of the IAF's post-1973 development program was to achieve both goals.

IAF commander Benny Peled considered the key challenge to be the acceleration of the location-strike cycle, which could not be achieved without entirely new intelligence methods that would bring the process toward the ideal of real-time intelligence. That would enhance IAF capabilities for all missions, not just the counter-SAM mission. A study of the efficacy of the costly IAF effort to assist the IDF ground forces in the 1973 war also concluded that insufficiently up-to-date target intelligence had been the number-one problem.[5] Soon enough it was determined that the key technological solution was to use remotely piloted vehicles (RPVs; now drones) small enough to avoid being hit, with a flight endurance long enough to maintain an almost constant presence over the battlefield, and with a video camera and datalink to enable real-time transmission of data to operational planners, who would no longer have to wait for aircraft to land to analyze photographs. It helped that Israel was a world leader in manufacturing increasingly capable RPVs.

The Drone Revolution

It was in Artzav 19 that drones first played a central role in combat operations, and they greatly exceeded expectations, marking the beginning of a new military era. Yet the air forces of the world did not exactly rush to acquire drones, let alone integrate them into their operations. Nine years later, in the US buildup for the 1991 Gulf War, the only observation drones in hand were imports from Israel procured by the Navy and Marine Corps. No other service had shown any interest in unmanned aircraft, and the US Army had canceled its own promising Aquila program in 1985 for dubious reasons, while the Air Force did not even start one.[6] That collective inaction was all the more remarkable because the Advanced Research Project Agency of the US Defense Department had successfully demonstrated a drone in 1972, just when the first Israeli drones were being tried. So the IDF were the first military forces anywhere in the world to have the great advantage of operating routinely with near real-time intelligence, a lead that proved surprisingly persistent.

Responding to the ever-growing threat of abundant and effective Soviet-made surface-to-air missiles during the War of Attrition, the Israeli Air Force began using Teledyne-Ryan 124 / BQM-34 Firebee target drones as SAM decoys, under the designation Mabat, the Hebrew acronym for "pilotless airplane."[7] Attempts were made also to attach cameras and send them to photograph areas deemed too dangerous for manned aircraft. In 1971, the air force acquired US-made Northrop BQM-74 Chukars, which it designated Telem, as noted with reference to their futile use on the Golan in 1973. The Telems were modified to follow a preprogrammed flight plan and enhanced electronically to simulate the radar cross-section of a much larger manned jet fighter, so as to awaken SAM batteries and radar controlled antiaircraft guns into action, thereby revealing their positions to attacking aircraft.[8] But using these drones for photographic reconnaissance proved less effective.

Shortly after the October 1973 war, after disappointing results with the converted US-made target drones, IDF Military Intelligence had begun using the Israeli-made RPV sold internationally as Mastiff, designated Sorek in the IDF. But it could not provide real-time coverage—it could only take still photographs the Intelligence Branch could use instead of requesting photographic missions by manned reconnaissance aircraft. Needing the real-time relay of the photographs, the air force attempted to retrofit a stabilized video camera with a datalink on a Chukar / Telem. It then turned to Israel Aircraft Industries (IAI), which responded quickly with the Scout (IDF name Zahavan), which could monitor large areas for many hours and relay photographs in real time to the analysts' video screens, to enable strike aircraft to take full advantage of the short exposure windows during which mobile SAM batteries were most vulnerable to attack. Other intelligence tools were also developed or procured, including airborne command-and-control systems, but the innovative use of drones was very much an Israeli advance.

Mastiff and Scout exemplified Israel's small, modestly funded defense industry at its best.[9] Their design was highly responsive to IDF needs because of the continuous intercommunication between the active-duty officers who wanted them and the engineers who were developing them, who were mostly IDF reserve officers. These RPVs were as simple as possible, mechanically robust, and designed for rough handling in field conditions. Because their design incorporated as many off-the-shelf components as possible, they were also cheap. The first models had television cameras and relayed imagery to the operators. Later, laser range finders were added to allow the use of the same drones by artillery spotters and to lase targets for manned aircraft.

At the start of what became the Scout project, IAI assigned a team of engineers to the task, including Yair Dobster.[10] He described the project as "basically a start-up. [Youngsters] with an adventurous spirit were recruited and an experienced team leader was appointed to guide us to tame the youngsters' tendency to wander too far as young, openminded, and fearless young people sometimes do." The firm treated Scout as if it were an ordinary manned aircraft; "it was made of the same type of aluminum with the same rivets, just like airliners are made to this day," according to Dobster. To save time and money, and to skip trial-and-fix iterations based on wind-tunnel experimentation, the proven double-boom design of the Arava light transport aircraft was simply downsized and paired with a single rear-facing engine. This design also made it easier to balance the weight distribution when operating with different payloads and/or extra fuel.

Designing the aircraft—the platform—was only the start. To operate the Scout, a ground-control station was needed with displays, joystick activators, and telecommunications. The optical payload also had to be developed from scratch to suit the *Scout,* and a mode of operation also had to be anticipated, there being no established operating doctrine.[11] One technical choice did not favor simplicity: in the United States the first experimental drone (which never reached serial production) had a fixed camera built in, with a floating-mirror apparatus to stabilize the image. That produced a reversed mirror image for the operator, adding a complication that would be undesirable if the system were ever used in real combat conditions. For Israelis that scenario was a given, and IAI decided to solve the problem by developing a gyrostabilized gimbal for the camera.

Much later, two members of the first crew of IDF operators described the initial growing pains, including part shortages that forced the cannibalization of some Scouts to keep others flying. Frequent, if small, engineering changes had to be made on the go, with no pause for verification by outside test and evaluation inspectors or for the attribution of blame for errors, as redesign decisions were made on the spot. It was a process that had no end date, with continual tinkering that did not end even when the Scouts were first delivered. All the while, the Scout's operating crews remained in close, informal contact with the developers, as is Israeli practice, to give them the feedback they needed. Initially there was much skepticism in the air force, which was apt to forget that drones existed at all. After sending in a debriefing report following a sortie that nearly ended in a crash, the drone squadron received phone calls from air force headquarters asking if the pilot was alive and well. Following that incident, every safety report from the drone squadron sardonically included the reassurance that "the pilot is alive and well."

In 1980, the first drone squadron participated in a divisional maneuver in the Sinai Peninsula. At the outset, the divisional commander, then BG Ehud Barak (later chief of staff, defense minister, and prime minister), told the air force drone unit commander that his unit was far down the priority list for the upcoming maneuver. But early in the morning, the squadron launched its drones with short-takeoff rocket boosters, only to discover that the pontoon bridge meant to simulate the crossing of the Suez Canal was being secretly moved to a new location by maneuver referees, in order to confuse Barak's division. When Barak noticed what was happening on his monitor, he insisted that the drone squadron fly nonstop until told otherwise; he had discovered the value of extended, real-time intelligence.[12]

The Scout squadron (the first of its kind anywhere) became fully operational in 1981, just when Syrian SAM batteries were deployed to the Beqaa Valley in Lebanon, thus extending their SAM coverage and threatening IAF operations in the northern front. There was also another wholly unexpected episode on May 14, 1981: one of the squadron's Mabat drones achieved a confirmed kill when a Syrian MiG-21 flew into the terrain while trying to shoot it down.[13] But it was their use by the IDF in the 1982 Lebanon war that really validated drones, starting a global race to develop them. Certainly, it was a real trial by fire as Scouts successfully hunted down highly mobile SAM batteries, including the then formidable SA-8s.[14] The first drone to succeed the Scout was its direct offspring, the IAI Searcher 2 drone, which entered service with the IAF in 1992 as the Khogla (Alectoris). By then the IDF were using drones to lase targets for precision-guided munitions.

The Palestinian offensive of 2000–2006 highlighted the need for more capable drones, with more endurance, which was amply provided by the large IAI Heron 1 (Shoval in the IDF). A new drone squadron was duly added to monitor the dense urban environment. Drone footage that revealed the true course of reported events also proved invaluable in the diplomatic and media arenas to support Israel's efforts to expose deceptive propaganda.[15]

What began as a search for a survivable reconnaissance aircraft evolved over time to the design of a new type of strike aircraft. The air force entered the era of armed drones with the Hermes 450 (the IDF Zik), which served as an attack drone with guided missiles to provide divisional headquarters with a dedicated platform for real-time intelligence, but could also be used to launch attacks on land or at sea.[16] By 2006 the IAF were operating a much bigger drone—the Heron TP (IAF Eitan), a much-enlarged derivative of the Heron that can attack ground targets at ranges in excess of 1,500 nautical miles (its one-way range is 4,000 nautical miles) with an endurance in excess of fifty hours. Hence it can advantageously replace manned aircraft for surveillance, reconnaissance, and also long-range attack. (Reportedly it has been used to interdict arms deliveries to Hezbollah deep inside Sudan.)[17] Certainly, its operational range suffices for air attacks with significant payloads anywhere in Iran.[18] A later addition to Israel's drone arsenal, the Hermes 900 (IAF Kochav), which became operational in 2015, can reportedly operate continuously for over twenty-four hours without refueling and carry up to four AGM-114 Hellfire missiles.[19]

A different type of Israeli-made strike drone was prominent in the 2020 Nagorno-Karabakh fighting. While other drones, specifically the Turkish Bayraktar, received more publicity, the IAI Harop, successor of the IAI Harpy attack drone or "loitering munition," seems to have had the greatest impact on the battlefield. Originally designed as an antiradar loitering munition, the canister-launched Harop has a mission endurance of up to nine hours and can serve for reconnaissance and area patrol, but also has a 16 kg warhead to dive on high-value targets. The Azeris used the Harop and two other Israeli models of loitering munitions, the Orbiter and the Skystriker, against all types of targets—SAM launchers, radars, tanks, APCs, artillery, infantry positions, and even trucks and buses used for troop transports, dissuading their use and thereby attacking both the morale and the mobility of their adversary.

Increasing Survivability

Avoiding the ground-launched missiles and gun danger zones required a no less precise knowledge of their location than was needed when striking them. However, if the targets were in the danger zones, evading missiles by flying low, as was attempted in the War of Attrition and the 1973 October war, brought the aircraft into the range of antiaircraft guns, whereas evading the guns by flying above their effective ceiling, as was also done during those wars, exposed the aircraft to missiles. Furthermore, new missile types, such as the man-portable SAM-7 (first employed en masse during the 1973 war), SAM-9 (a vehicle-mounted version of the SAM-7), and SAM-8 (which along with the SAM-9 arrived during the 1970s), could hit aircraft at very low altitudes. During both wars the IAF finally chose to usually attack above the gun ceiling and within the SAM envelope, employing a mix of surprise, decoys, and electronic countermeasures (ECMs), but mostly relying on aerobatic flying and complex teamwork to reduce losses. The IAF also attempted various methods of semistandoff attacks with Shrike antiradar missiles, electro-optical Walleye gliding bombs, and HOBO gliding bombs, and the very fast medium-altitude toss-bombing (Kela) technique to launch munitions and egress before the antiaircraft missiles reached the aircraft. But all of these proved to be technologically deficient, and the targets were missed all too often.

By contrast, a real advance was using unmanned aircraft as decoys. Flying before or during air strikes, the drones caused the enemy to waste time and ammunition on the wrong targets while simultaneously exposing themselves to discovery. The decoys used in the 1973 war had proved their great potential, even though the opportunity to capitalize on their effectiveness had been missed. But having learned their value the IAF broadened its arsenal of decoys. Newly developed, unpowered Shimshon (Samson) gliding decoys, along with older Telem decoys, were launched by the attacking aircraft in the 1982 takedown, successfully luring the Syrians to reveal the positions of their missile batteries and to expend missiles uselessly.[20]

Improved ECMs were obviously important to protect aircraft flying within the range of enemy SAMs. One of the problems revealed by the 1973 war had been that the available ECMs were fairly effective against older SAM-2s, only slightly effective against the later SAM-3s, and totally ineffective against the latest SAM-6s. Evidently the US ECMs effort was too slow to keep up with the rapid pace of Soviet SAM innovations. It was time for local efforts. On the second day of the 1973 war a burnt-out SAM-6 seeker head had been retrieved on the Golan Heights. Another still-intact seeker was retrieved on October 24 from the Suez Canal front. Both were sent to Rafael, whose countermeasures team immediately started working at a furious pace to develop specific jamming and deception ECMs against the SAM-6, then the most effective Soviet antiaircraft missile.

Those electronic countermeasure devices were designed, engineered, tested, and manufactured in a few months instead of years, and were delivered to the air force in the spring of 1974—too late for the October 1973 war but ready for the next.[21] At the same time, old US-made ECM pods that had failed in 1973 were modified with some new electronics, and they too turned out to be highly effective against the SAM-6s. However, the constant threat was that the enemy would acquire a new system impervious to the existing ECMs. Maintaining viable ECMs requires a constant effort to collect information on the latest enemy systems, find their limitations, and then develop the capabilities required to jam or confuse them. As with the Israeli drone industry, Israel invested heavily in this field and still does.

An even better solution is not to have to fly into the danger zone at all, but that requires munitions that can reach targets from beyond the enemy's effective interception range. Despite all the disappointments with the standoff munitions it first received from the United States, the IAF invested heavily in this field, purchasing various US-made electro-optically guided antiradiation missiles (collectively called Egrof, "Fist") with a color suffix to indicate the specific munition, for example, Egrof Yarok ("Green Fist") for the GBU-15, Egrof Khum ("Brown Fist") for the Israeli Tadmit, Egrof Tzahov ("Yellow Fist") for the modified AGM-62 Walleye, and so on.[22]

By 1982 the air force had integrated AGM-78 Standard antiradiation missiles (code-named Egrof Sagol, "Purple Fist"), which were technically superior to the AGM-45 Shrike received a decade earlier, both in their range and because they were programmed to keep flying toward the targeted radar even if their operators switched them off. Previously, it had been enough for radar operators to briefly stop radar emissions to deprive the AGM-45's homing guidance and to restart them when the radar dish had rotated to another bearing, to divert the attack. But with the AGM-78 this tactic would fail because the missile was programmed to continue on its initial track, so the entire radar would have to be moved—impossible to do in seconds. The AGM-78 was so effective that the Sixty-Ninth F-4 Phantom Squadron was specifically dedicated to employ them, with air crews trained in their use; eventually the Sixty-Ninth would launch some thirty AGM-78s during Operation Artzav 19 in 1982.

The air force also availed itself of the proximity of the SAMs to Israeli territory—it would not be flying across the seas to a distant target, as all the targets were just up the road. To add to the small number of expensive air-launched variants, both the AGM-45 and AGM-78 air-launched missiles were drastically modified into ground-launched missiles. This started with the mounting of AGM-45s on World War II–vintage M3 half-tracks, to obtain an 11 km-range system introduced at the end of the 1973 October war as a stopgap measure. Later their range was increased by adding booster rockets, which had been developed and manufactured locally and very quickly. The boosted AGM-45 Kilshon ("Pitchfork"), developed and tested within two weeks, used converted M-4 Sherman tanks as the launching platforms, yet another use for that forty-year-old mainstay; the missiles could attack targets up to sixty kilometers away.

In 1977 the more capable AGM-78 Keres ("Hook"), introduced with a more elaborate truck-mounted triple launcher, had a longer range and interim inertial guidance to hit SAM radars whose operators had stopped emitting between launches, precisely to throw off antiradiation munitions. But this was a case of a rapid, economical, and seemingly clever innovation that failed in combat. While dozens of Kilshon and Keres missiles were fired at Syrian SAM batteries, they failed to destroy any; evidently with ground-launched missiles the initial trajectory angles were just too flat. It was fortunate for the air force that the air-launched radar-killers proved sufficient. Also essential, it was determined, were electro-optical guided munitions, which allowed aircraft to deliver their munitions accurately while remaining beyond the range of the Soviet 23 mm antiaircraft cannons and portable infrared missiles with which the Syrians were amply supplied. (There have never been any Western antiaircraft cannon even remotely as cost effective as the Soviet 23 mm in its twin or quad configurations.)

The air force employed a mix of homegrown and US-supplied guided munitions that could be launched from some distance from the target. What the air force called the loitering attack method was optimal for standoff weapons, of which it had a useful variety: the US-made AGM-62 Walleye and the GBU-8 HOBOS were older guided but unpowered bombs, which glided down to their targets within modest standoff ranges.[23] The Tadmit, locally developed by

Rafael, was also a glide bomb manually guided from the launching aircraft.[24] Zeev Bonen, then Rafael's CEO, was well aware of the pressing need for standoff munitions and ordered the conversion of one of the company's production lines to manufacture Tadmit standoff munitions exclusively, supplying the IAF with the first unit as early as the end of 1974.[25] This marks Tadmit as a precursor to Iron Dome: it too was developed very quickly by ignoring normal procurement, development, and manufacturing procedures and practices to provide a rapid solution to a major threat.

A small fraction of Tadmit bombs were directed to their targets by weapons officers on board C-130 aircraft that were thought better suited to accurately deliver the bombs to their targets than the F-4 Phantoms, in which the descending glide bomb had to be observed through a tiny cathode-ray tube. To train crews in the use of the new electro-optical munitions, a US simulator was employed at Eglin Air Force Base in Florida from 1978, with the program extending into 1982.[26]

The Orchestra—Sella's Computer Revolution

All the capabilities accumulated would not suffice if the entire operation could not be precisely coordinated from start to finish. The various capabilities had to be combined in a common integrated action plan that would enable all the myriad parts to work together in exactly the correct sequence and with precise timing. But as the envisaged IAF operation grew in size and complexity, its preplanned coordination in the manner of a well-rehearsed orchestra would impose increasing rigidities in the plan implementation, evoking fears of a repeat of the Dugman and Tagar failures.

Acquiring real-time intelligence with a rapid planning cycle to exploit it would only partially reduce the time lapse between acquisition of the targets and strikes against them. If the pilots and aircraft waited for orders on the ground and needed to study them in depth before takeoff, there would still be a considerable delay. Therefore, the best solution was to have aircraft already in the air waiting for targets that would be provided in such a way that the pilot did not need much time to study and implement the orders. But doing this with up to a couple of hundred aircraft in the air would require a constantly updated picture of the overall situation of both enemy and friendly forces.

When the British invented the centralized aerial battle control system that had saved them in the summer of 1940, the control was done manually. The IAF, established and initially manned mostly by veterans of the Royal Air Force, adopted the same system, with the commander in charge sitting on a balcony overlooking a large room containing a big table with a large-scale map of the Middle East on which women conscripts manually moved around small tags, each representing an aircraft with all its details (type, armament, fuel state, current altitude., current speed) manually written on it.[27] Until 1973 this method had proven successful, but it became clear it could not keep up with the much faster tempo and far greater complexity of operations. The IAF needed new capabilities, a new plan to exploit them, and a new way of commanding the operations.

On the morning of June 9, 1982, a few hours before Operation Artzav 19 started, Colonel Aviem Sella, then head of the operations branch directly under air force chief MG David Ivry, was in his underground headquarters awaiting the moment he had been preparing for ever since the 1973 war. "There are many different components to this story," Sella later said, "and their common denominator is that they all rose from the largest manufacturer of motivation—failure . . . the air force had been insulted."[28] Born in 1946, in the air force since 1963, Sella flew Israel's first F-4E Phantoms during the War of Attrition, shooting down five MiG-21s, including one flown by a Soviet pilot in the ambush operation of July 30, 1970. Sella was not mentally prepared for the 1973 defeat inflicted by the SAMs.

Very soon after the fighting ceased in 1974, young Sella was assigned to IAF headquarters with the rank of major and given the large task of finding ways of suppressing the Arab air defenses that had proved so formidable. Chief of Operations Amos Amir formed six different teams to think through as many aspects of the overall problem: electronic warfare, intelligence, training, ordnance, and more. Sella moved among them, sometimes to listen, sometimes to lead. Though only a major (hardly a senior position in the IDF, with its few stars), Sella found himself in charge of the single most important air force initiative.

One of Sella's first and arguably hardest tasks was to change the IAF mindset that saw strong opposition of many pilots to the very idea of focusing on fighting the SAMs.[29] The Old Guard still thought only of air battles—the test of quick instincts, endurance, thorough knowledge of the aircraft on both sides and of their limits, with a willingness to push those limits and take risks—all of which did indeed yield the air combat superiority Arab air forces could not overcome, and which caused them to rely so largely on Soviet-supplied air defenses. What the Old Guard could not accept was that it was their own superiority as pilots that had driven the other side to rely on the missile defenses that had defeated the air force in 1973—hence it was not a problem that more air combat superiority could possibly cure. In the immediate aftermath, another group kept arguing that the defeat had been caused by the sequence of disastrous last-minute reversals in attack priorities, so that the solution for next time was to stick to the plan and add more self-protection for more combat aircraft rather than diverting air force funding to missiles, drones, and supporting aircraft.

Sella, himself a deputy squadron commander during the 1973 war, immersed himself in the subject and won over the senior officers for his ideas. He presented them in an internal 1975 document, "Missile Combat—Aerial Warfare against SAM Batteries," which developed a previous publication on the subject by Eytan Ben-Elyahu, an F-4 Phantom squadron leader who would go on to become IAF chief. Sella further promoted the idea that anti-SAM combat warranted the same methodological approach and resources as air-to-air combat, laying out a detailed breakdown of the solution: first, avoid detection via proper flight-profile planning (normally ultralow entry); second, break the SAM radar lock-on with maneuvering and radar-confusing chaff; and, third, disable the SAM radar with electronic countermeasures and skilled use of radar-warning receivers and jammers. In addition, air-crew awareness and skills should be raised with SAM battery scale models, illustrations, and even full-size mockups at every IAF air base, so airmen could practice identification-and-attack runs every time they flew their landing approach. He also suggested establishing special ranges with simulated SAMs, a very costly training aid.

To implement Sella's vision, the air force had to add another layer of command-and-control capability; plans were still based on the meticulous centralized planning that had been so successful with Moked in 1967 but had failed in 1973 when the plans could not be adapted to changing circumstances. The new layer allowed the central air command center to adapt or change plans even when the aircraft were already in the air in the midst of operations. A new five-step workflow emerged under Sella's leadership:

1. Strike formations go on loitering routes in front of but out of range of the enemy's SAM array.
2. A specialized intelligence team acquires and relays real-time intelligence on the array of SAMs, recording their movements and pinpointing their positions.
3. That team transfers its synthesis, backed up by aerial photographs, to the anti-SAM command post.
4. The latter relays the position of each SAM battery to the loitering aircraft best positioned to attack that particular battery.
5. The aircraft then launch standoff electro-optical munitions at the SAM batteries, which are vectored to strike their fire-control centers.

The command post established to coordinate the successive attacks against the SAM batteries would have to command and control flights of up to 200 aircraft at the same time, in addition to land-based platforms and electronic warfare assets. Having designed the new plan and the command post that would execute it, Sella was appointed to direct its lean staff, comprising an intelligence officer, an air traffic control officer, and a specialized planning officer for each one of the three SAM regiments of the Syrian air defense network, as well as an electronic warfare officer. Once established in the IAF's underground command headquarters, the staff attempted to implement Sella's five-step process in a trial run against a mobile SAM battery. The test failed, and that failure pointed to the urgent need to computerize the entire workflow, hardly an everyday challenge when only the most routine processes were computerized.[30]

The tightly scheduled attack plan, with all the different ground as well as air units involved—from drones, decoys, and helicopters to the large number of fighter-bombers—could not be coordinated and controlled with the old manual methods. Parameters could have been memorized, but there were just too many variables. Moreover, once the operation started, updated instructions for personnel and machines would be needed in seconds—much too fast for instant human recalculations of the entire strike plans.

Computer control was essential because the planners had to anticipate that three or four attack waves would be needed to destroy the vast, varied array of Syrian air defenses, each flown by aircraft armed with different weapons to attack each one of their separate components (radars, missiles, launchers, command posts, mobile antiaircraft cannons, and towed antiaircraft gun mounts). They had to simultaneously process the locations, weaponloads, and fuel status of all air force aircraft, and the nature and location of all the targets, thereby allowing continuous optimization of the attack by matching aircraft to targets.[31] Any delay would create a dangerous air traffic problem, diminish the element of surprise, and expose dozens of attacking aircraft to air defenses. By then mainframe computers had been in standard use in all modern countries for almost two decades, but there was no standard program, or set of programs, that was at all suitable to command and control such a complicated plan of operations. Furthermore, the cost estimate for a tailor-made program (it was all hand coding then) killed the idea of computerizing the IAF's command and control.[32]

Ironically it was again the ace fighter pilot, Major Aviem Sella, who initiated the effort to acquire a digital command-and-control system, named Periscope; the air commanders in their deep bunker would "see" the air battle through it. Having studied computer science during his prescribed midcareer university education leave, Sella was convinced he knew everything there was to know about computers. Armed with the cocky attitude of a typical fighter pilot, Sella went directly to air force chief MG Benny Peled to tell him that the air force had to be computerized. This occurred in 1974, many years before email or Google, when computers were still seen as merely computational machines, not the core of operating systems. Hence, Sella was told that the computerization of air operations was neither necessary nor possible. Undeterred, he went on to search for ways to realize his vision. What he needed was a program that could integrate and continuously update in near real time all essential data into the ongoing operational plan, such as the exact location of a just-moved Syrian missile launcher or the weaponload of a specific fighter at a given moment.[33] It would all become very ordinary later on, at least for the US armed forces and a few other advanced armed forces, but at the time it was certainly a macroinnovation without precedent.

He received some good advice when he witnessed an artillery command exercise and asked Amnon Yogev, a reserve artillery officer working at the Weizmann Institute, how his branch coped with the challenge of directing the simultaneous firing of many artillery guns and rockets against a large number of targets of varying types, many mobile. Yogev referred Sella to Zvi Lapidot, a director in the institute's Computer Science Department and a reserve signals officer of an artillery battalion, who was working on a computerized command project for the Artillery Corps. Sella requested a meeting with the president of the Weizmann Institute (an august figure in Israel), and promptly convinced him to assign a team of computer scientists to work under his direction to develop an integrated operational system for the air force. One was duly formed and set to work, and they were not Weizmann's rejects but rather the A-team. The Weizmann Institute Computer Science Department had acquired its first computer in the 1950s, when even telephones were scarce in Israel, and had developed an advanced capability in that field.

Sella had not changed the minds of his more than skeptical superiors; he had simply gone ahead, never asking for nor receiving authorization, let alone a budget, from the air force. The lack of any money was not an obstacle to starting the project, as the scientists simply remained on the Weizmann Institute's payroll. After six months of hard work, the prototype program for the intended system was ready. Sella went to the air force chief, Peled, and persuaded him to visit the institute with him to "see something." One day in the summer of 1975 Peled arrived at the Weizmann Institute, just when the power was cut off. Peled did not storm off but waited patiently for the power to be restored. He spent two hours examining the system before declaring, "We need this, as it is, by tomorrow."[34] Within a week, a truck had come to take the bulky mainframe computer from the Weizmann Institute to the air force headquarters bunker. There were no formalities, no paperwork, no bills to pay—it was just a matter of unloading the computer from the truck.

As compared to the manual display board of 1973, Periscope belonged to a different era. It integrated, instantaneously, the action of individual systems—for example, a single fighter-bomber—into a supersystem orchestrated centrally that could follow up initial strikes with ad hoc strike orders for loitering aircraft hunting for mobile surface-to-air missiles. It could do so on the basis of continuously updated information received from drones, ground radars, airborne command centers, individual fighter aircraft, and more. The most dangerous SAM batteries were the mobile ones that could move every ten minutes—much too fast for the command structure and intelligence of the air force of 1973. But in 1982 the Syrian SAM network was faced with an air force that could redirect its aircraft, weapons, countermeasures, and decoys within seconds.[35]

The Periscope system induced a complete mentality change in the air force, and rather quickly. With that, the air force's internal organizational structure also changed. The system did not really centralize everything—its development generated the realization that it is neither possible nor desirable to command and control all air combat, all close air support sorties, and all reconnaissance and transport missions, as well as to coordinate the fight against the SAMs, all from one and the same control center.[36]

Considerable training was required to enable all the components of the IAF capable of implementing the new concept. A full-scale operable model of a SAM-6 battery was set up in Hatzor Air Base, and there were models of Soviet early warning radar at every IAF air base, also at an electronic aerial range. David Ivry, the air force commander, decided that each new fighter pilot was to perform at least one practice run against Syrian SAM-2 and SAM-3 batteries in South Lebanon, but without any firing. That way every pilot in the designated anti-SAM squadrons would be familiar with the theater of operations and the tactics developed to destroy SAM batteries. In addition, every four months, the IAF conducted a large-scale exercise centered on SAM suppression. Those exercises were nicknamed Torpedoes and included extensive use of simulated Syrian antiaircraft units, and later of simulated combat over Lebanon against the actual SAM batteries, without the Syrians becoming aware of what the IAF was up to.

Action

As to what actually happened between 16:00 and 18:00 on June 9, 1982, even now the specifics remain secret, perhaps because of a single technical detail.[37] But there is no doubt that the day started with reconnaissance and electronic-intelligence sorties, backed by ground and air-launched decoys that duly aroused Syrian SAM batteries, revealing their positions. All known SAM positions were relayed to the air force command center and the parameters were loaded into the Periscope program.

Scout drones then verified the SAMs' location and status prior to attack, with four electronic warfare aircraft sent up to provide barrage jamming against Syrian radars.[38] At that point the fighter-bombers of the strike force went up to hold loitering positions at different altitudes, with F-15s and F-16s armed with air-to-air weapons flying top cover to fight off any Syrian fighters that tried to intervene against the stacked fighter-bombers. The strike force consisted of twenty-four F-4 Phantoms armed with antiradiation and electro-optical precision-guided munitions, supplemented by A-4 Skyhawks and Israeli-made Kfir fighter-bombers armed with both cluster and ordinary bombs.[39] Their presence was precautionary; in case the Syrians jammed the antiradiation missiles or employed some unknown Soviet countermeasure against the electro-optical munitions, the A-4s and Kfirs would resort to classic dive-bombing.[40]

At the peak of the operation, a hundred Israeli aircraft were in the air, armed with different munitions for different missions, so that whenever an attack plan for any particular set of targets was generated, based on information processed by the Periscope program, an optimized strike package could be made up by selecting aircraft from those already in the air and waiting. The hostile and friendly aerial picture and aerial traffic control was provided by Israel's array of ground-based radars, their operators talking directly to the pilots. A year before the war these had been reinforced with the arrival of E-2C Hawkeye airborne radars employed as forward air-traffic controllers for the loitering fighters, to keep their various formations properly stacked until each had its turn to launch an attack, based on their proximity to a verified SAM battery location. The E-2Cs also served to extend the range of radar coverage beyond the reach of Israel's radars high on Mount Hermon and could relay radio communications if needed.[41]

As the attack unfolded, whenever a Syrian radar was switched on, AGM-78s would be launched, usually destroying it. When a visual or radar contact revealed the location of a SAM battery, it was promptly attacked by one or more F-4s armed with Tadmit or GBU-15 remote-controlled gliding bombs, usually aimed at the battery's fire-control center. A-4 Skyhawks and Kfirs with cluster bombs would follow up to destroy the launchers of "headless" batteries.

During the two hours of Artzav 19, the Syrian SAM batteries in Lebanon remained stationary, so that the much-practiced ability to hunt mobile batteries was not needed. But in the following days, when SAM batteries, including SA-8s, moved at night into South Lebanon they were destroyed within hours.[42] It was only then that the drones fully realized their potential—during Artzav 19 they had mostly served to verify the validity of battery location data just prior to attack, to make sure the batteries had not suddenly moved or been destroyed.[43] But during the night of June 9 the Scout drones played a crucial role in hunting the formidable SA-8s. The latter were found alongside the other SAM batteries the Syrians sent into South Lebanon under cover of the night; with Periscope guidance and real-time imagery from the drones, the command post directed the "shooters," the airborne fighter-bombers, to destroy the moving batteries, providing the pilots with updated and exact locations.

As head of the IAF's operation branch, Sella had the privilege of seeing the conversion of his concept into an operational plan, Artzav 19, as well as its actual implementation in war, using the computer-assisted command system he himself had introduced into the air force just a few years earlier.[44] It was, in fact, the first computerized war operation attempted anywhere. It was also the first time the air force deviated from its much-valued unified unitary command and control directly by the air force commander, because of the expectation that important events would happen at a frantic pace in warfare compressed to just a few hours, leaving no time for staff deliberations. Instead, air force commander MG David Ivry limited himself to supervising air-to-air operations and appointed Sella to directly command the takedown of Syrian air defenses by overseeing the functioning of Periscope.[45]

A visitor to the underground air force command center would have been greeted by the odd sight of an ultraorthodox Jew in a typical black suit and hat: Menachem Kraus, the only member of the Weizmann Institute team who actually knew how to operate the mainframe computer on which the Periscope software ran. Having never served in the IDF (exempted as a full-time cleric), Kraus did not even have a common soldier's minimum-security clearance, let alone the more demanding one needed for the ultrasecret command center, and Sella had to convince his chief that Kraus was critical for mission success. Sella later recounted that while the operation was in full motion, every few minutes Kraus raised his hand and turned his fingers to signal how many batteries destroyed while Sella, sitting across the command room, would raise his fingers in response to inform him of the number destroyed up to that moment. When it was all over, Kraus crossed the corridor to Sella's office to shake hands, covering his eyes to avoid having to look at all the young women soldiers along the way (in the summer heat the air conditioning could not keep up at full occupancy, so dress was informal).[46] Artzav 19 in June 1982 was a watershed event in the history of warfare. It was the first battle fought under computer command for all practical purposes; it was also the first battle, aerial or otherwise, in which unmanned aerial vehicles, or drones, had a major, arguably decisive, role.

[12]

Elite Units

The Mass Production of Military Excellence

HISTORICALLY, LARGE AND WELL-ORGANIZED ARMED FORCES HAVE HAD little use for special or commando units. During the twentieth century when long wars were fought by large conscript forces, in which even a small country like Israel fielded more than 200,000 soldiers (in 1973), the major armies had to raise, train, equip, and deploy thousands of company-sized combat units of around a hundred soldiers, each of which needed some hard-fighting men to lead the rest in battle. Hence almost all senior military officers opposed the skimming of the best men to form elite units.

Political leaders in pursuit of military glamor to relieve the gloom of long wars might press for the establishment of commando or special units—Winston Churchill was a notable enthusiast—but army chiefs were typically reluctant to devote attention or resources to establishing a few small commando or otherwise special units, which would always be too small to win battles on their own, too valuable to be wasted on mere skirmishes, and too hard to integrate usefully in large-scale operations. Enthusiasts would argue against this by offering variants of the Trojan War ruse, in which a small number of heroes finally made the victory of the Greeks possible by jumping out of the horse to open the gates of the walled city, still today the most famous of history's force multipliers.

But most twentieth-century army chiefs were unimpressed, because they saw great obstacles to brilliant Trojan horse operations in real life, including the difficulty of coordinating the actions of small special units with the movements of large regular forces; had the Greek army crept up to the gates and Helen detected the ruse, they would have been massacred by the Trojans above them on the walls. Classically trained or not, American, British, Soviet, and German army chiefs all preferred to do without special units. Their priority was to keep the best fighters in regular line units, so that the few could energize the many. It was only the end of large-scale wars that allowed the present proliferation of special forces of all kinds.

The founding father of the IDF and Israel's first prime minister, David Ben-Gurion, had been the first to realize that the new state would need an army just to survive from its very first day, and it would have to be a large army. For exactly the same reason as the professionally trained generals, he therefore opposed elite forces that would deprive units of essential leaders. He would have wanted to abolish the elite force of the War of Independence, the Palmach, in spite of its epic achievements, even if it had not been politicized by its leaders as it was. And once victory was secure, he did abolish the Palmach's headquarters. Hence the postwar IDF started off in 1949 with no elite forces but for a single and small British-style paratrooper battalion of very modest attainments, and a small if highly effective unit of sea commandos who had sunk Egypt's flagship the year before. As minister of defense Ben-Gurion was content that it should be so: he wanted good brigades of thousands, not exceptional platoons of thirty.

As of the early 2020s, by contrast, the IDF have a wide variety of special operations units, a transformation that started in a small way when Ben-Gurion was still minister of defense, not because of any change in policy but as an urgent response to an immediate problem. Eventually, other specific threats emerged against which large-scale responses were inappropriate and induced the establishment of more special operations units, a process favored by the country's strategic circumstances: the last major war against regular forces occurred in 1982 in Lebanon, whereas fighting against irregular enemies has continued unabated, waxing and waning in intensity, the type of war that suits special units.[1] Now the IDF have an entire array of different elite units that are territorially specialized, for the verdant north, the arid south, the Red Sea–Eilat area, and the highly urbanized center of the country. They are also functionally specialized, for intelligence infiltration, long-range reconnaissance, long-range strike and undercover operations, tunnel warfare, and more, in addition to three larger top-tier units that are more versatile while retaining different core specializations.

It all started with Unit 101. In the aftermath of the War of Independence the newly victorious IDF was in dissolution as veterans returned to their homes, when infiltrators started crossing the undemarcated and unfenced armistice lines into Israeli territory, sometimes only to harvest their own lost fields, but sometimes to steal, rob, and kill. Sizable gangs came for both plunder and revenge—from summer 1949 to the end of 1956 there were approximately 11,500 such attacks on Israeli civilians and their property.

None of Israel's Arab neighbors policed their side of the Green Line (named for the color of the 1949 armistice lines on IDF military maps), and neither could the IDF disperse its few soldiers all along the border, where they could not train for war. The inordinate length of the armistice lines as compared to the country's exiguous total territory, the result of Israel's long and narrow shape, made it impossible to outpost and patrol the borders usefully.[2] The War of Independence had ended in the spring of 1949 with armistices, not peace treaties, because in the politics of the defeated neighbors it was axiomatic that war would be resumed as soon as there was any chance of success, with cross-border infiltration attacks welcomed as a token of what was to come on a much larger scale—most conducted by Palestinians, but many also by various state armies.

This meant that Israel's military leaders had to confront two very different military threats: the first was clearly the threat to fundamental security (*Bitachon Yesodi*) with major offensives intended to defeat the IDF, then physically annihilate the Israeli state and its Jewish population. To repel that threat Israel would of course need a modern army, with artillery, tanks, an air force, and more. The second was the threat to routine security (*Bitachon Shotef*), with sniping, small ambushes, mines, raids to kill Israelis, and thefts, whose overall aim was to wear down the resolve of the Jewish population to live in Israel.

To counter the everyday threat, a particular kind of military strength would be needed, best suited for "small wars." That immediately posed a dilemma for Israeli military planning, because the requirements of large-scale war are qualitatively different.[3] The dilemma became more acute as the routine security situation continued to deteriorate: in 1950, 67 Israeli citizens were killed by infiltrators; in 1951, the number rose to 137 killed or wounded;

and in 1952 casualties rose to a combined total of 182. All the victims were civilians, mostly women and children, with a total of 1,751 incidents in 1952.[4]

The initial Israeli response was defensive, including more border patrols and more ambushes, but there was no possibility of protecting even thinly the disproportionately long and meandering armistice lines. Protests to the Mixed Armistice Commission were to avail. The fifty-odd UN observers compiled reports but could do little else. Clearly, the government had to protect its citizens. After failing to achieve any results with diplomatic means, it decided to use force to compel the neighboring states to control their side of the border, but this had to be force well short of all-out war. First came a classic public warning from Prime Minister David Ben-Gurion that "if the armistice lines along the border are open to terrorists and murderers . . . we reserve our freedom of action."[5] Internally, Chief of Staff Moshe Dayan explained the reasoning: "We must determine the rules of what is and is not allowed in our relations with the Arab Countries, and we must be careful not to be submissive and acceptant of [attacks] against us, even if their [effect is small]."[6]

Dayan was an expert when it came to reprisal raids. As a youngster he had learned the trade from a master of the craft as a prized recruit in British captain Orde Wingate's Special Night Squads, raiding offending Arab villages during the 1936–1939 uprising. In World War II Dayan became a scout for the British Army leading raids into Vichy French Lebanon, leading the way for a British force on June 7, 1941, that penetrated Lebanon, losing an eye in the fighting. As chief of staff, Dayan logically opted for commando operations: raids to attack targets behind enemy lines that would rely on surprise and guile instead of numbers or heavy firepower, and that would clearly be understood as reactions to attacks within Israel. In so doing, Dayan formulated a policy and security concept that Ben-Gurion accepted, thereby institutionalizing such retaliatory actions within enemy territory. At first, however, an attempt was made to use airpower: on April 5, 1951, eight fighters attacked Syrian Army outposts in the southern Golan at El Hama in the Yarmuq Valley. The government was surprised by the vehement reactions of British, French, and US diplomats (these days ground actions evoke stronger protests than air strikes), and Ben-Gurion decided then and there to stop relying on airpower.[7]

But when it came to ground attacks, Dayan and the general staff soon discovered that the troops they had just could not fight. After the 1947–1949 war most of the best fighting officers had left the IDF to pursue their civilian lives. Ready as they were to return in uniform if the country had to be defended in another big war, they were unwilling to serve in peacetime as career officers. The best forces, the Palmach brigades, had been disbanded by Ben-Gurion and not replaced, while the line units were filled with new immigrants who rarely knew enough Hebrew to understand their orders and whose basic training had been hurried. Morale and discipline were so low that in 1951 not a single battalion was deemed truly ready for combat.[8]

In 1950, Dayan, as head of Southern Command, had been sorely disappointed by the performance of the Seventh Armored Brigade under his command. After the Jordanians suddenly claimed that the road to Eilat infringed on their territory, they blocked passage for several days. Dayan had immediately given orders to the brigade to clear the road forcibly, but was very displeased by the "dithering, indecisive way" in which his orders were implemented.[9] The next year, as the new head of the General Staff Branch, he oversaw a much worse failure: on May 2, 1951, a force of Syrian village militia and regular infantry entered the Israeli side of the demilitarized zone along the border and occupied Tel Mutilla, a small rocky hill just north of Lake Tiberias. The IDF Northern Command reacted immediately by sending an infantry unit to repel the Syrians. But repeated Israeli attacks were inept and easily repulsed. More troops had to be sent, and it took five days of fighting to finally drive out the Syrians, which required the participation of the Druze battalion.[10] By the end, forty IDF soldiers had been killed and seventy-two wounded. The duration of the fighting and the casualties were out of all proportion to the small size of the Syrian force, which according to Syrian press reports suffered some 200 killed—the majority of the force. It was a clear sign something was badly amiss with the infantry's training and morale, a message reinforced by a string of failed reprisal raids at Wadi Fukin, Beit Sira, Beit Awwa, and Idna, where on January 25, 1953, two companies of the nominally elite 890th Paratrooper Battalion were repelled by enemy fire, failed to mount enough suppressive fire, and retreated without completing the mission.

But the most humiliating failure occurred on January 23, 1953, when an infantry battalion of the once-famed Givati Brigade was ordered to launch a night raid against the central Jordanian village of Falame, a notorious nest of transborder marauders. Built on a hilltop, the village was not the easiest of targets, but it was only defended by a dozen riflemen of the Jordanian

National Guard with no heavy weapons. The Israelis repeatedly lost their way in the dark; when the advance unit finally reached the edge of the village it was met by scattered rifle fire. When six Israeli soldiers were wounded, the battalion commander ordered a retreat back into Israeli territory. Three senior officers, including Moshe Dayan, were waiting for the return of the battalion just across the armistice line to hear the raid's results. Dayan's first reaction was to discharge the commander on the spot, but then he realized that other officers and other units would not have fought any better.[11]

In 1953, out of eighty-five military operations, forty-six were outright failures, and only fifteen were deemed successful. It often happened that would-be raiders turned back because they could not find their objective in the night. On some occasions they were repulsed, and officers greatly exaggerated enemy strength in their reports.[12] Successive failures further reduced the already-low morale of the troops. It was obvious that the night-fighting skills and sharp combat edge of the War of Independence had been lost. In one incident, a platoon sent to blow up a well in the Gaza Strip lost its way and failed to locate its objective in the darkness. When morning came, the Israelis discovered that they had wandered in circles without even having crossed the armistice line.[13] Deeply frustrated, Dayan noted in his diary that "even our elite units that were trained for special actions such as the parachute brigade, exhibited shameful negligence resulting in many of our actions ending up as failures."[14]

It was a vicious cycle: because the policy of retaliatory raids was undermined by the army's inadequacies, Arab infiltration increased, as did the death toll of civilians and soldiers alike. That in turn demoralized the civilian population, already enduring severe economic hardships because Israel's thin economy had been overwhelmed by the mass arrival of destitute immigrants. Food rationing and low incomes were all the harder to bear when the new Jewish state could not even protect its population from daily attacks. Under great pressure, Dayan bitterly evoked the disgraceful failure at Falame to rule that no officer would be permitted to suspend an attack once it had commenced unless it had suffered a casualty rate in excess of 50 percent. That was exceedingly harsh in an army whose guiding principle was to reduce casualties to the absolute minimum.

By then it was clear that the army's melancholy state could not be remedied by exhortations or regulations. Officers lacked confidence in their men, many of whom were unmilitary new immigrants, while the men were demoralized by their officers' visible reluctance to rely on them in battle. And each new failure caused a further decline in the Army's self-confidence.[15] It was Colonel Michael Shaham, commander of the Jerusalem Brigade, who came up with a possible remedy: the formation of a small force of skilled and dedicated fighters who could be counted upon to carry out reprisal raids with determination.[16] Shaham's preferred recruits were junior officers who had done well in the war but had resigned in 1949 rather than stay in the peacetime army. And he argued that the unit he advocated would have to remain outside the army's formal structure with all its rules, because "the right kind of man" would not volunteer to return in uniform if it meant accepting the discipline of a peacetime army.

In August 1953, Chief of Staff Mordechai Makleff decided to accept Shaham's scheme, overruling Dayan's objection to the idea that the army would rely on small special units to do what he believed every combat unit should be able to do.[17] Makleff chose Ariel Sharon, a twenty-five-year-old civilian student in the School of Oriental Studies at Hebrew University, to lead the special unit. He had fought well in the War of Independence and held the rank of major in a reserve battalion.[18] Dayan already knew and respected Sharon; they had served together under Northern Command in 1952, when Sharon was a military intelligence officer while Dayan was head of the operational branch. One day Dayan asked Sharon to look into the possibility of capturing two Jordanian soldiers to be used as bargaining chips for two Israeli soldiers who had strayed into Jordanian territory and were held by the Arab Legion. Sharon was noncommittal but with another officer he drove straight up to the border, where he saw and captured two legionnaires at pistol point whom he brought back to Dayan, who later said: "I asked him if it was possible, and he returned with two Arab Legion soldiers as if he had gone to pick fruit in the garden."[19]

Upon receiving the go-ahead for the creation of his Unit 101, Sharon handpicked his men one by one, traveling around the country to persuade aggressive and intelligent fighters he knew personally or by reputation to abandon civilian life and join him in hard combat.[20] Unit 101 never included more than forty-five men at any one time, but they were all excellent fighters—fighters rather than soldiers, for they wore neither uniforms nor badges of rank, and their weapons were not standard issue. Some of Sharon's men proved excellent tacticians who would later advance to high rank; one who did not, Meir Har-Zion, soon became a legendary figure, "the best soldier

Israel ever had," Dayan once said. When Har-Zion's sister was murdered by Bedouin tribesmen while illegally hiking across Jordanian-controlled territory with her boyfriend to visit the Petra monuments, Har-Zion crossed the border with some friends, tracked down the Bedouins responsible, raided their village, and killed four in revenge, as is customary among the Bedouins themselves.[21]

Not yet chief of staff, Dayan was initially opposed to the elite-unit solution, but became an enthusiastic advocate after meeting Har-Zion and his comrades. He could see that they would revive the dormant skills of the Army of Independence: superior fieldcraft to infiltrate enemy territory and night fighting. He therefore hoped they would both devise new tactics and uplift the IDF's morale by successful actions. The five months of its existence, from August 1953 to January 1954, proved sufficient for the aims of Unit 101. Its last operation was also the largest and most controversial. On October 14, 1953, after the murder of a woman and her two children in a village near Lod Airport in the heart of Israel, the unit was sent to attack the strongly held Jordanian village of Qibya.[22] Sixty-three men of the army's paratrooper battalion were sent as a covering force for Unit 101's forty, who fought their way into the village, rounded up the inhabitants, and blew up forty-five houses in retaliation. But not all the houses had been fully evacuated beforehand, and some forty villagers still hiding within were buried under the rubble, resulting in a total of sixty-six dead and seventy-five wounded civilians, including those hit during the battle. The brutality of the raid led to sharp protests in Israel and abroad, and the result was a sharp policy change: future retaliation raids would be aimed at military units, not villages. "Israel has learned," Dayan declared, "that even when the Arabs hit civilian population, we must aim at military targets."[23]

From 1953 the overall strategic situation was changing because Egypt's new dictator, Gamal Abdel Nasser, decided to start a confrontation with Israel using Palestinians to do the actual fighting. By early 1954, the Egyptian Army was raising units of fedayeen (Islamic self-sacrificers) in Gaza to infiltrate Israeli lines to attack civilians. That put pressure on King Hussein of Jordan to show that his army also supported fedayeen raids. In response, the Israeli government decided to attack military bases providing training and support to fedayeen units. Attacking military installations obviously increased escalation risks, but in 1953 the decision seemed inevitable because neither passivity nor continued attacks against border villages were feasible options.

Because the aim was to influence leaders, the operations had to be larger in scope and intensity than the earlier village attacks.[24] That in turn meant that the original Unit 101 format was much too small, while a larger force could not possibly remain so informal; it would need a proper organization. Dayan decided to merge Unit 101 with the paratrooper battalion instead of disbanding it, which was duly done one month after he became chief of staff on December 7, 1953. Thus by January 1954 Sharon became commander of the combined Unit 101–890th Paratrooper Battalion, a force ten times larger than the original Unit 101. Intensively trained commando forces in the modern style have existed since 1916 (for example, the Sturmtruppen of Imperial Germany) but Israel's innovation was to use them to raise fighting standards across the entire army, first by direct expansion as Unit 101 inspired the much-larger paratrooper battalion, which then expanded into the still-larger Paratrooper Brigade, more than a fiftyfold increase. That took care of Dayan's original reservation (shared by army chiefs everywhere) that special forces would weaken much-larger regular forces by taking away their best men—the precious handful of really good fighters who can energize entire units. Moreover, young officers formed in the expanding elite force were distributed to lead units throughout the IDF.

At the start of this process, Sharon faced the challenge of combining the most informal IDF unit with the paratroopers, the IDF's best parade-ground performers. To start with, most paratroop officers requested and obtained transfers to other units rather than serve under Sharon, who promptly appointed an officer who remained as his deputy. This was a fortunate choice, because Aharon Davidi would himself earn fame as a fighter and leader. (He rose to chief officer of infantry and paratroopers.) As adaptation and bonding proceeded, the paratroopers learned to scout and fight in small groups while former 101 men learned to operate on a larger scale and use heavier weapons. The old paratroopers soon lost their spic-and-span appearance even as they were becoming warriors in the relentless Unit 101 style. It was characteristic of the times that the battalion's "field exercises" were actually guerrilla-style raids into enemy territory.[25]

The 202nd Paratrooper Battalion, as it was called, demonstrated its new skills in a March 28, 1954, night raid against the fortified Jordanian village of Nahalin, some ten kilometers west of Bethlehem, in response to a massacre in the Negev in which eleven Israeli bus passengers were killed and others wounded. After brushing aside the local National Guard defenders, the

Israelis blew up a set number of houses but, mindful of the Qibya deaths, they carefully inspected the buildings to ensure they were empty. There were eight more raids by the end of 1954, all successful.[26]

With that the IDF acquired a reliable military instrument, one that could keep growing in size even while setting undiminished standards for the rest of the army. The process was facilitated by the role of parachute-jump training in the battalion: it was treated more or less as a sport to uphold its distinctive esprit de corps but was unlikely to be used in war on a large scale. As it expanded, new recruits would only require a few qualification jumps, not an expensive program of multiple day and night airdrops.

The 202nd soon started innovating tactically. Its new technique for attacks against fortified positions replaced the IDF's version of the British "fire and movement" two-step, whereby one unit acts as a stationary fire team shooting at the enemy to keep their heads down, while another acts as the assault team that advances, before the two switch roles for the next move forward.[27] In July 1954 Sharon was wounded while leading a raid against an Egyptian stronghold facing the border kibbutz of Kissufim.[28] Like many others along the armistice line, it consisted of concentric trench lines linked by narrow communication trenches, with barbed-wire fences and mines all around the perimeter. While lying in his hospital bed, Sharon worked out a new tactical method to replace the old British tactics.[29] Instead of relying on heavy covering fire, the men were to approach the trench system without firing at all. Walking slowly and in absolute silence until fired upon, the men were then to run forward as quickly as possible, firing on the move, while the barbed-wire fences were breached by Bangalore torpedoes, long metal tubes filled with high explosives. Once they reached the trench line, the men were to form small assault groups. Without pausing to clear the fire trenches, they were to jump into the communication trenches, running and shooting all the way to the center of the stronghold and then out again. This way the force would sweep trench line after trench line. Under this method, the assault teams were to keep moving and shooting until all the defenders were killed or captured.

The essence of Sharon's new tactic was to exploit the shock effect of sudden attack followed by relentless advance, intended to first surprise the enemy and then to break his will to resist, rather than to win by killing as many as possible. But his method was especially vulnerable to enemy counterattacks: while the paratroopers were fighting inside the stronghold, but not yet in full control, the arrival of enemy forces on the scene could catch them mid-stride and easily disrupt a tactic that relied so heavily on morale effects, by rallying the defenders and overwhelming the scattered teams of attackers.

Because the 202nd Battalion frequently fought at night and its men were trained to fire on the move, the paratroopers were mostly armed with 9 mm submachine guns firing pistol ammunition. The Israeli-made Uzi was a good weapon that had been adopted by several foreign armies, but no submachine gun is accurate beyond a hundred yards, and only a few men can hit even large targets while firing from the hip. Instead, because the Uzi with its rapid fire is deadly at close quarters, Sharon's men were taught to close the distance as quickly as possible and go for hand-to-hand combat distances. Egyptian and Jordanian soldiers tended to fight poorly at close quarters, though they were often good riflemen and quite a few Jordanian legionnaires were real marksmen. By attacking at night, when accurate long-range fire was hindered if not impossible, the paratroopers deprived the Arabs of their advantage, while benefiting from their discomfort in night fighting.

Few of Sharon's original officers survived unscathed the reprisal operations of the mid-1950s. First, the actions were of course perilous for everybody, and second, "follow-me" leadership increased unit effectiveness but also put commanders at higher risk than their men. Sharon himself was wounded, as were almost all his officers, some repeatedly, and others were killed. (Among the small number of survivors were three future generals: Mordechai Gur, Yitzhak Hoffi, and Rafael Eitan.)

Leading from the front persisted in spite of the loss of officers in minor operations. The wisdom of allowing officers with General Staff potential to fight and die in minor skirmishes has been repeatedly debated in the IDF, but it remains official doctrine. In theory, there is a cost / benefit calculation whereby the loss of much-valued officers is offset by the overall gain in combat morale in general and combat momentum especially. But in truth the "follow-me" ethos has such a strong grip on the mentality of the IDF that it has been very difficult to restrain even the most senior officers from placing themselves in the line of fire, even if they cannot actually take away combat leadership from their juniors.[30]

Dayan kept his eye on the paratroopers even after his appointment as chief of staff in December 1953 and was often at the sendoff point when they went into action. He wanted—and got—officers who were fighting men rather than managers in uniform. Moreover he wanted teeth with an absolute minimum of logistic tail—overdoing it more than once—and his reorganized paratroopers provided an exemplary model. To spread the paratrooper spirit in the army as a whole, Dayan insisted that every officer, including himself, undergo paratrooper jump training. He also enlarged the paratrooper unit so that by 1956 it was the size of a brigade. The other brigades responded to the paratroopers' prestige by trying to compete with them, which they could do more easily when they fought alongside each other, which the Golani First Infantry Brigade and Nahal infantry units started doing more and more. Ever since Unit 101 was raised, Sharon's men had monopolized combat missions. But when combat missions were assigned to ordinary infantry forces, they performed much better than before. When Israel went to war in the 1956 Sinai Campaign the transformation was complete: in less than four years since the shameful Falame debacle, Dayan's adage "Better to be engaged in restraining the noble stallion than in prodding the reluctant ox" applied in full, as soldiers in thinly armored half-tracks charged Egyptian positions as if they were riding in well-armored battle tanks.

The Unit 101 ethos persists in today's top-tier special operations battalions and the specialized elite units that are set up as needed and readily disbanded when no longer needed, a process obscured by the persistence of their evocative names in entirely new units established later on. Examples include the Shimshon ("Samson") Unit 367 of Southern Command that operated in Arab disguise in the Gaza Strip and was disbanded while its sister unit Duvdevan ("Cherry") Unit 217 in the West Bank persists, now with TV fame via its *Fauda* evocation. Likewise Egoz ("Walnut"), Northern Command's counterinfiltration unit active in 1963–1973, and the Rimon ("Pomegranate") counterinsurgency unit were disbanded in the late 1970s, but a new Egoz Unit 621 serves in Northern Command, and a new Rimon unit was established in 2010 as a desert-warfare scout unit within the Maglan (Ibis) long-range scouting unit, itself part of the Oz special operations brigade. The Shaked ("Almond") and Haruv ("Carob") reconnaissance battalions were dismantled in the mid-1970s and reincarnated as ordinary infantry battalions in the Givati and Lion brigades, respectively. The overall idea is to pursue the advantages of specialization without organizational rigidity: if a unit no longer fits current needs, it is simply abolished.

IDF special units differ from each other, but overall, while they superficially resemble the commando forces of other armed forces, the similarity is deceptive.

Elite special operations units around the world that achieve high standards are manned by experienced career soldiers who first serve in line units. The IDF, by contrast, relies on young conscripts for all its combat forces, including its special operations units. Competition to serve in elite units upon enlistment is fierce among young high school students—only admission to air force flight training is equal in prestige. To improve their selection chances, many young Israelis join special prearmy preparation programs to upgrade their physical fitness and learn army ways, with the more affluent even hiring veterans as personal trainers in some cases, an interesting variation on the pastimes of wealthy youths elsewhere.

Once future conscripts inform the IDF of their desire to volunteer for a special unit and are found to have the minimum physical and mental requirements, they are summoned to an army base for a "Yom Sayarot," a Special Forces (testing) Day. It amounts to a series of physical and mental tests to determine who among thousands of seventeen-year-olds qualifies for the further admission tests of the IDF's top-tier elite units: the Sayeret Matkal of the Intelligence Corps; the naval commando and frogman unit Shayetet 13; the air force Sayeret Shaldag Unit 5101; and the Airborne Rescue and Evacuation Unit 669. Those who pass are sent for another week of physical and mental tests (*Gibush*—"tryout"), with the highest scorers sent to the Gibush Matkal for the top-tier units. Psychological character and stability tests are integral to the process because soldiers in special operations elite units not only face severe challenges, as do all soldiers, but are also much more likely to be on their own, or near enough.

The initial training of the conscripts admitted to one of the top-tier elite units lasts roughly twenty-two months, and it is the longest initial training course in the IDF, except for air force pilots and naval officers (whose training includes higher education for an academic degree). The training contents include infantry staples: fieldcraft, individual weapon training, physical hardening by unarmed combat, and simple combat simulations, diverging from the US / British / French norm only in omitting parading and saluting, while adding a particular IDF emphasis on very long marches in the Wingate tradition. Advanced individual training varies with each unit but invariably includes field navigation in different terrains, counterinsurgency basics, air-to-ground cooperation, airborne operations, intelligence gathering, sharpshooter instruction, medic training, and more.

The constant inflow of high-quality, enthusiastic youths from all parts of society energizes the IDF in general and the special units especially. But one obvious disadvantage of manning special forces with young recruits, no more than twenty years old when fully trained, is their lack of experience as compared to their counterparts in other armies, notably the British SAS, the US Army's airborne detachment Delta, the US Navy SEALs, the US Special Forces Green Berets, or the French RIPMA, who almost invariably start training in those units only after several years of prior service in line units. It seems, however, that the lack of prior experience is offset by the sheer frequency with which IDF special units are sent into action, so what remains are the advantages of the young volunteers in their intellect, personality, and even leadership charisma. They are, after all, not self-selected for military careers as noncommissioned officers, as is the case for nonofficers in US and other special forces. On the contrary, they expect to emerge as the future business, professional, academic, and political leaders of the country. (As of 2021 both the outgoing and the incoming prime ministers had served as junior officers in very demanding special units, committing to additional years of service beyond the compulsory three years in order to do so.)

That virtue, however, is also a potential problem, and a familiar one: the concentration of talented youths in small special units deprives the IDF line forces of good squad leaders and good sergeants, and also deprives the IDF as a whole of good candidates for officers' school. That follows from the fact that the IDF have no military academy—all officers rise from the ranks. To safeguard the leadership potential of special-unit recruits, many are sent to officers' school upon finishing their unit training program, or else later on; when they graduate from officers' school, few return to their original elite units, with most sent to mechanized infantry, armor, or other line units to command platoons at first or to take up other staff and command slots. Those who do remain in the IDF after both their compulsory and signed-on added service to become career officers, must usually add to their special operations training by going through training in armor or artillery or signals or some other branch, and then also learn how to command and control such forces.[31]

A noted example of such a career path is the late Lieutenant Colonel Yonatan Netanyahu, who famously commanded the top-tier Sayeret Matkal commandos during the July 1976 Entebbe hostage rescue (in which he was the only IDF soldier killed). At one point, he had commanded a tank battalion, after going through armor training, starting with gunnery, driving, and maintenance skills. Another example is Ehud Barak, who also started his service in the Sayeret Matkal, and also went through armor training and became a tank-battalion commander before going on to staff and command courses, eventually becoming IDF chief of staff and later prime minister.[32] The movement of officers in and out of the special units extends across the regular line forces—it is not unusual for special units to have officers from the Golani and Nahal infantry brigades, Thirty-Fifth Paratroopers, and Givati infantry brigade, and line officers can even serve as unit commanders. One example is Moshe "Boogie" Ya'alon, who became IDF chief of staff (2002–2005) and afterwards defense minister (2013–2016) after starting as a recruit in the Fiftieth Paratrooper Battalion before eventually becoming the commander of the Sayeret Matkal, with more training for further promotions including command of an armored division.

Conscripts who volunteer and are accepted by one of the elite units are required to sign up for extra time on top of their mandatory service. Service in the four top-tier units, Sayeret Matkal, Shaldag, Shayetet 13, and the Airborne Combat Rescue and Evacuation Unit 669, requires an added thirty-six months of paid career service in addition to the thirty-two months of compulsory service, amounting to five years and eight months in uniform, which can become eight years to complete a university degree while still in uniform. Even service in second-tier elite units requires at least an extra year of paid career service, sometimes two, which means that their intake can only start studying or working postarmy at age twenty-three or twenty-four, years behind their American or European counterparts. That should have severe consequences for Israel's economy as well as society, but it seems that the all-round competence acquired during military service, and the leisure to pursue private studies, go a long way.

The top-tier special operations units have the privilege of being first in picking new recruits from the conscript intake. Recruits not selected by any of the top-tier units can go on to try their luck with other special units, but in practice many conscripts prefer to join a tier-two unit, for any number of reasons: they have a brother or a father who served in it before them, have friends serving there, have heard of a unit's accomplishments in a recent operation, or believe the specific character of the unit fits their personality. One such unit, the undercover Duvdevan counterinsurgency specialists who operate in Arab disguise, has the acquired glamor of the internationally successful *Fauda* TV series, no doubt attracting more recruits. Another, Maglan

Unit 212 for deep rear intrusions and actions in distant places, appeals to would-be explorers, while the current Egoz, specializing in operations in the wooded terrain of the north, has the acquired importance of opposing Hezbollah, Israel's most active antagonist.

For many years these diverse units were also very independent, under the loose supervision of the Chief Infantry Officer and the operational control of one of the three regional commands for the North, Center, and South. But in 2015 they were all placed under the single headquarters of the newly formed Eighty-Ninth Commando Brigade.[33] It was hoped that the brigade could function as a cohesive combat force when required, while still retaining the special expertise and ethos of each of its different units.

Certain larger formations not officially considered special operations forces nevertheless have elite status not only in popular opinion, but also in the effectiveness calculations of war planners. Among them are the Thirty-Fifth Paratrooper Brigade, which is actually a light infantry brigade not really meant for airborne assault by parachute, though it treasures its red caps and boots; the Golani First Infantry Brigade, which dates back to the birth of the IDF, having started in 1948 as a Haganah light infantry brigade of foot soldiers but is now equipped with the heaviest armored infantry fighting vehicle in the world, the sixty-metric-ton Namer; the revived post-1984 Givati Brigade originally trained for amphibious warfare; and the 933rd Nahal Brigade, from the Hebrew acronym for "pioneer fighting youth," originally manned by conscripts from collective farms (kibbutzim and moshavim) and youth movements, who combined combatant service with agricultural work. Its Fiftieth Airborne Battalion has a famous combat record. Finally there are the elite special-purpose forces that also attract volunteers, such as the Yahalom of the Engineering Corps, which focuses on high-tech tunnel warfare in addition to its tense IED and ordnance disposal duties, alongside more prosaic obstacle breaching and demolitions; and the top-tier Air Force 669 Combat Search and Rescue unit, which is continuously active because in peacetime it conducts urgent civilian medical transfers.

The IDF also has a number of even more specialized, or rather localized, reserve units that come into existence only when mobilized for combat operations or when recalled for refresher training. One is the Alpinist Unit, the 7810th Reconnaissance Battalion trained for the snowy slopes and ice peaks of Mount Hermon; when local snow is insufficient, the unit is sent to train in the Alps. Another is Lotar Eilat, recruited in that Red Sea town as a quick-reaction, local counterinfiltration and hostage-rescue force.

In the special forces even more than in the line forces, the reserve units are kept together by small numbers of the highly committed, who have been training together for years and tend to socialize with each other in between their duty stints. While their in-group mentality can be a problem in cooperating with other units, they do offer high levels of commitment and proficiency, with performance standards almost as high as those of the active-duty forces, and in some things even higher, while the budgetary cost of maintaining such units is of course much lower. The special units also function as experimental, or beta, organizations for the IDF as a whole by being the first to try new weapons and tactics, which they can do more easily on their small scale instead of waiting for more resources for larger line formations. Hence the special units act as the spearhead of IDF innovation, both technological and tactical. Their informality and particular culture of continuous learning facilitate innovation and generate tech-savvy entrepreneurial types who now play a large role in the country's technology-driven economy.[34]

Recent examples are the tactics developed by the Yahalom unit of the Engineering Corps to penetrate and crack Hezbollah "nature reserves," IDF slang for the underground strongholds of bunkers and tunnels found in rural areas of southern Lebanon.[35] Using a variety of location, penetration, and attack techniques, this "nature reserve" package has been disseminated from Yahalom to the line brigades that would be sent into action in the event of a full-scale fight. Likewise, during the 2014 Protective Edge operation in Gaza, Yahalom had to learn—in the heat of battle—how to locate and destroy Hamas offensive tunnels, before the advent of today's high-tech equipment. Again, those techniques were quickly disseminated to other units.[36]

As with any large organization, the IDF can miss optimization opportunities because of obstacles to communication between specialized units, especially because many operate under secrecy rules. In response to this, the latest addition to the long list of IDF special units was established with the mission of overcoming those barriers. This "Ghost Unit" was formed by Chief of Staff LG Aviv Kochavi in 2020 to function as an elite but nonspecialized combat unit whose personnel are collectively familiar with the capabilities and limitations of all IDF components, and whose mission is to select and integrate the most relevant capabilities for any given combat task at hand. In other words, unlike the others this elite unit cannot be self-absorbed but must instead keep scanning the military horizon within the IDF in order to perform its mission. Operating in every Area Command—North, Center, and South—and in every dimension—land, sea, air, and also underground—the unit's mission is to bring to bear everything the IDF has to offer at any one point in time to execute the mission at hand.

Kochavi realized that the IDF has a wealth of advanced capabilities but no effective counter to the age-old problem of organizational suboptimization. An IDF infantry squad in the field that finds itself in a skirmish with a similar enemy force would have to fight on equal terms, gun versus gun, obtaining no benefit from the altogether larger capabilities of the IDF as a whole, tantalizingly present in theory, absent in practice.[37] That is the challenge the Ghost Unit is to address: how to bring to bear the full might of the IDF to magnify the strength of whatever units are engaged in combat at any one time. Because of its mission, the unit does not recruit and train its own, instead receiving seasoned teams from different units with different specializations such as the undercover Duvdevan unit, Yahalom elite combat engineers, the Oketz canine unit, and more. The basic requirement is a combination of good fighting skills with technology savvy.[38]

Each combat team is matched as needed with support staff from intelligence, cyber, and air units and also with civilian professionals such as engineers and computer experts. By definition the unit is multiservice and multidimensional. For the Ghost Unit the challenge is to overcome organizational, interunit, and interservice procedural barriers when striving to integrate and optimize air and land firepower with the maneuver elements. A first remedy was to form the so-called Sufa ("Storm") teams with the varied skills and capabilities needed to process large amounts of data, in order to precisely orchestrate strikes from various ranges from land, sea, and air platforms. The Golani Brigade, which serves in the north facing Hezbollah forces that can attack at any time, was the first to include Sufa teams within its forces. Another initiative advanced by the Ghost Unit is the use of drones and micro-UAVs in urban terrain, always difficult for any kind of combat operation anywhere, but especially so for the IDF because of their double casualty constraint (enemy casualties being politically costly) and the prevalence of high-density housing in hostile areas.[39]

It is not yet clear if the Ghost Unit will persist and mature, or if it will reach a culminating point in its development to be replaced by another attempt at optimization, because the IDF need not prioritize the upkeep of hallowed traditions over optimization, having other sources of motivation, chiefly the recurrence of immediate danger. The use of functioning combat units as—in effect—experimental laboratories can be traced back to Unit 101, Israel's first commando unit that was both small and ephemeral and yet widely and enduringly influential.

[13]

Military Entrepreneurs and Special Forces

UNIT 101 WAS ESTABLISHED BECAUSE THE PRIME MINISTER AND THE IDF hierarchy headed by the chief of staff, faced with a demoralized postwar army, decided that even if very small, a truly effective combat unit could achieve morale-raising tactical victories that would in turn allow more victories, with widening effects across the army. But the other top-tier IDF special operations units resemble today's start-ups: they originated in the initiative of an individual officer who had identified a significant capability not adequately provided by any existing IDF force, but which could be provided with the right concept, training, and structure. Obviously, it would take an original mind to get that far, but to actually create the missing combat unit another thing was required that does not easily combine with a lively intellect: sheer persistence. Even though IDF headquarters were very small by world standards, they were large enough to resist new and untested ideas that claimed already-scarce resources. Those who had that rare combination of originality and dogged persistence became the founders.

The first, as we have seen, was the serial innovator and future naval chief Yohai Ben-Nun, who founded the IDF's Shayetet 13 naval commandos, personally determining their doctrine, choosing their equipment, and defining their instruction program in every detail, with himself as chief instructor at first. But perhaps the most successful IDF military entrepreneur was the later Brigadier General Avraham Arnan, founder of Unit 269, later Sayeret Matkal (reconnaissance unit of the General Staff), often known simply as "the unit" (*Hayehida*), designed for intelligence-gathering ground actions behind enemy lines (sometimes so far behind that they end up in another country entirely) under the direct command of the IDF Intelligence Branch. Later, more specialized units were started for such operations as hostage rescue, but until then as the IDF's only elite commando unit (once Unit 101 was dissolved), the Sayeret Matkal did it all, from hostage rescue to long-range strikes.

When IDF headquarters were first established on the classic model, the General Staff included a G-2 intelligence section to study enemy capabilities and enemy intentions. It received generic information from foreign ministry diplomats, too generic mostly, as well as from press sources, the reports of border patrols and observation posts, and a very small Unit 154 that sent agents in disguise across the border, that is, covert agents that looked like locals, mostly Arabs, as opposed to clandestine intruders hiding in nature and at night.

Techniques to tap into enemy telegraph and telephone lines date back to 1914, and the IDF did have simple listening devices it could plant on wires in enemy territory. One such device in Syrian territory could intercept military communications quite reliably, but now and then a team had to sneak across the border to replace the batteries. In December 1954 a mixed team of five soldiers crossed the border to reach a device deep in Syrian territory, but it was ambushed by a larger force and surrendered. The Israelis were held in separate cells for three weeks and severely tortured. One (Uri Ilan) committed suicide, leaving behind a note that would later become famous: "I did not betray." The senior man, Sergeant Meir Yakobi, decided to emulate Samson by leading his Syrian captors to the concealed device to then activate its self-destruction charge and kill them and himself, but heavy rains had incapacitated the electrical trigger.[1] In March 1956, after fifteen months of captivity, the four survivors were released in exchange for thirty-five captured Syrian soldiers.

What was learned from that failure was that clandestine operations entailing the risk of capture and interrogation could not be left to ordinary infantrymen, however brave and capable. That was the simple starting point of a young major serving in Unit 154: Avraham Arnan, a former Palmach platoon commander—one of the few who had become professional officers instead of joining the majority in civilian life who served only when recalled for reserve duties.[2] Because Arnan's experience was that Unit 154's Arab agents were not very productive—they were mostly rural folk with little access to intel sources—he advocated their replacement by ultratrained soldiers who might also be good infantrymen or good commandos for assaults, but who would be selected and trained to serve as deep-penetration scouts, skilled in clandestine (hidden, not disguised) infiltration, and technically expert handlers of advanced listening devices and varied sensors.[3] More ambitious operations would follow, but in the first instance the unit Arnan envisioned was to carry out short incursions into hostile countries to gather immediate and accessible intelligence. That seems modest enough, but Arnan had to persist mightily to turn his ideas into reality. The IDF General Staff just had no time for him, focused as it was on raising large combat forces for the large war it feared was coming, which was in fact imminent.

It was only after victory in the 1956 Sinai Campaign, with senior officers more relaxed, that Arnan started building his force by recruiting some fellow veterans from the Palmach's "Arab unit" of Arabic-speaking Jews who looked like Arabs, such as Iraqi and Yemenite Jews. Actually, Arnan's clandestine unit was itself built in clandestine fashion because Arnan used Unit 154 facilities without any explicit authorization from superior officers.[4] That ensured both the cheerful informality of Unit 101 and also its acute logistic shortcomings—with more equipment "organized," that is, pilfered, than properly issued.[5]

Without any official permission from the head of the Intelligence Corps or the IDF chief of staff, Arnan nevertheless proceeded to look for recruits with suitable personalities and skills across the army. He would then rely on sheer effrontery and his mysterious Unit 154 credentials to obtain their reassignment to his own nameless unit. For training, he moved his chosen men through existing paratrooper and other courses before bringing them back to his unit within the Intelligence Corps, in spite of the intense skepticism of the commanding officer, the quietly thoughtful, indeed scholarly, Major General Yehoshafat Harkabi. It was Harkabi's skepticism—a great virtue in an intelligence chief—that saved the unit, as that skepticism extended to his own negative view of Arnan's project, which he therefore did not stop.

Arnan persisted on his own until a mid-1958 General Staff decision finally authorized his clandestine effort to organize a superior clandestine force, but only on a tiny scale bound to be inadequate. This Unit 269 was placed directly under the head of the Intelligence Corps, a clear token of importance, but its authorized strength was only fourteen in all.[6] That number was much too small to build up different subunits of any effective force, but Arnan was advised by that earlier military entrepreneur Ariel Sharon to "first say yes" and then to push for more billets until he reached his own target number.[7] Arnan finally began to organize the unit he had dreamed of, which was to be "totally different from what existed before it, a unit that did not follow any route previously trodden, and that would be able to think and implement in a way no one has thought about before."[8] That claim to originality would be validated by future operations, but in one respect Arnan was perfectly conventional: his men had to be perfectionists when it came to the different skills needed, including highly accurate shooting as riflemen, snipers, machine gunners, and also as users of handguns. Those very basic skills saved the day on more than one occasion.

Though he actually wanted to build something quite different, Arnan thought it useful to claim the legacy of Unit 101, and therefore tried to persuade its veterans to join him; his most valued recruit was the legendary Meir Har-Zion.[9] The ploy worked, and the best recruits soon wanted to join the excitingly secretive 269, which could finally grow because the commando-minded IDF chief of staff Moshe Dayan became actively interested in Arnan's enterprise and gave him leave to expand it.[10] Arnan would go around military bases looking for the right stuff, trusting his intuition rather than records or tests, and acting with a great sense of urgency, with good reason: Dayan's tenure was coming to an end.[11]

Dayan's successor as chief of staff, Haim Laskov, was a British Army veteran.[12] In contrast to the impetuous rule-breaker Dayan, who had personally led daring raids behind enemy lines in 1941 and again in 1948, Laskov had risen to the rank of major in the British Army by meeting its standards in disciplined skills and orderly procedures. Predictably, he believed that the IDF needed more discipline across the board rather than more Unit 101 daredevilry. Acting as the entrepreneur he was, Arnan maneuvered for Laskov's favor by exaggerating the influence of the British desert-raiding Special Air Service started in 1941 by David Stirling on his own actually very different unit, even adopting its "Who dares wins" motto.[13] While Laskov was not entirely convinced, he did not try to stop Arnan's gathering of men, equipment, and facilities. Always a canny political operator, Arnan housed his unit not too far from the IDF's general headquarters in the heart of Tel Aviv, giving him easy access to top officers and allowing him to invite passing ministers to drop in to view the exacting and exciting training of his men—even Prime Minister David Ben-Gurion was an interested visitor.

As his unit grew, Arnan needed more funds, which the Intelligence Corps could not provide from its tight budget; its research unit was housed in an ex-British Army corrugated steel hut as late as 1970. Arnan became an institutional fundraiser by lobbying IDF generals, including future chief of staff David Elazar; but it was not enough, so his men became adept thieves of equipment and supplies issued to other units in other camps.[14] This "requisitioning" became an attribute of the unit during its early years and was even viewed as good training, for it required planning, stealth, and quick execution.[15] Arnan's constant pressure to field-test his unit eventually resulted in its selection for a first operational mission in early 1959.[16] The task was to infiltrate the Syrian-held Golan Heights to install intelligence gear. The mission was a success, and Arnan immediately secured another mission in Jordan that was also a success. The IDF's top officers were concerned about the freewheeling style of Arnan's men but pleased with its results.[17]

The charismatic Arnan evoked something of a personality cult around him that raised the morale of his men while doing nothing to enhance their respect for IDF rules, which their commander was himself outmaneuvering.[18] What finally convinced the General Staff to suppress its doubts and embrace Arnan's unit was the Rotem Affair, which erupted on February 20, 1960. The IDF was utterly surprised by the sudden entry of one Egyptian armored division and three infantry brigades into the Sinai. That was an abrupt, inherently very dangerous shift in the balance of military power on Israel's most important front, which Intelligence had failed to predict in spite of all the preparations the entry must have required and did not even monitor in real time. (By then the Egyptian Army was receiving sound Soviet training on how to move large armored forces in stealthy fashion by detailed preparations followed by very fast driveouts.) It was only four days after Egyptian armored columns arrived in Jebel Libni on the central Sinai axis leading to Israeli lines that General Headquarters found out, and that by mere chance: a two-division force had penetrated to face Israel's Negev border before the IDF could mobilize its reserve forces to hold them.

As it happened, the newly arrived Egyptian forces withdrew without incident but the impact on the IDF's generals was both immediate and lasting, validating Arnan's strongest argument for his unit.[19] Things changed for the better under a new chief of IDF Intelligence appointed in 1962, MG Meir

Amit, who increasingly relied on the unit to gather intelligence, at first only near Israel's borders, but later also deep in enemy territory.[20] As its successes increased, Arnan hosted parties for his unit to which senior officers were also invited, including his commanding officer, Amit.[21]

Because the unit's long-distance penetrations increased in both frequency and depth, reaching points very far from Israeli territory, Arnan reached out to Uri Yarom, the commander of the IDF's first helicopter squadron, Unit 124.[22] At that point, helicopters were seen as marginal by the fighter pilots in charge of the air force, because they had no defined combat role, only rescue missions and such.[23] It was only in 1956 that two Sikorsky S-55s were bought from the United States, despite opposition from the fighter jocks who would rather have bought one more Mirage; they reached the air force a few months before the 1956 Sinai Campaign.[24] In his autobiography, Yarom wrote that the air force disliked helicopters because they were viewed just like camels, only useful to move people about—but camels whose maintenance was especially expensive.[25]

As for the ground officers, they much preferred fixed-wing transports (like the ageless DC-3 / Dakotas) that could drop paratroopers in much bigger numbers, much farther, rather faster, and far more cheaply. But Yarom viewed the helicopter as a Trojan horse that could join the battle by circling round to land troops in the less defended, even undefended, enemy rear. Arnan wanted to find a way to send his men in deep-penetration missions to distant targets and pick them out again, ruling out parachuting, and he landed on the helicopter as his solution. He started joint training with Yarom's Unit 124, which culminated in a series of exercises that in turn led to the formulation of a joint combat doctrine.[26]

By 1962 ambitious deep-penetration missions were underway, for which pilots had to master new skills such as navigation on completely dark nights, landing in unfamiliar terrain, and avoiding radar detection by ultralow flying.[27] In 1963 and into 1964 repeated operations demonstrated mature capabilities: Arnan's men were landed by helicopters deep into Egyptian territory and picked up again several hours later, after planting devices to tap Egyptian military communications. By then it was becoming clear that the helicopter-plus-commando combination could generate powerful synergies even for large-scale combat operations (as in the 1967 battle of Abu Agheila, when Egyptian artillery batteries were attacked by helicopter-borne paratroopers) essentially because helicopters could evade radar detection by flying very low; their noise, though loud overhead, did not actually travel very far. With this modus operandi, commandos gathered much intelligence on Arab air bases for the Moked operation at the outset of the 1967 Six-Day War.

Arnan left his unit's command in 1964 to serve in intelligence headquarters, receiving promotion to the rank of brigadier general, a very high rank in the star-poor IDF.[28] Under its formal name Sayeret Matkal Arnan's creation continued to perform its intelligence role, often far from Israel's borders, while acquiring new roles, including counterinsurgency in response to the post-1967 upsurge. It would receive much more publicity than it ever wanted after Operation Isotope, conducted in full view of press cameras at Israel's central airport (then Lod, now Ben-Gurion): the rescue of the ninety passengers and ten crewmembers of Sabena flight 571 from Vienna to Tel Aviv that was seized on May 8, 1972, by four hijackers who demanded the release of 315 prisoners, threatening to blow up the aircraft. With no specific training for such a rescue, but with advice from former unit members turned air marshals for El Al flights (a widely copied Israeli innovation), sixteen Sayeret Matkal men, including future prime minister Benjamin Netanyahu, under the command of another future prime minister, Ehud Barak, approached in white technician overalls, ostensibly to repair the plane. They stormed the aircraft, relying on their excellent handgun skills, killing two hijackers and capturing the other two, with three passengers wounded, including one who died of her injuries (Netanyahu was also shot through his arm by a unit member).[29]

Much more spectacular was another commando-type action, the April 9–10, 1973, Operation Aviv Ne'urim ("Spring of Youth"), the assassination of three Palestine Liberation Organization (PLO) chiefs in Beirut in retaliation for the massacre of Israeli athletes at the 1972 Munich Summer Olympics, in which they were personally implicated: Muhammad Youssef al-Najjar of Black September, directly responsible for the massacre; Kamal Adwan, chief of operations; and Kamal Nasser, PLO leader and spokesman. Under commander Ehud Barak, who was disguised as a woman for the occasion (he had practiced walking in full disguise in Tel Aviv), Sayeret Matkal men arrived by boat at a Lebanese beach whence Mossad agents in three rented cars drove them to the two adjacent luxury apartment houses where the PLO chiefs lived. Three teams broke into the apartments and killed the targets, while three others led by Barak remained outside to fight off any PLO reinforcements or Lebanese gendarmes. With their highly accurate shooting, Barak and his men fought them off, killing a dozen before all returned to the cars that took them to the beach for a motorboat trip to patrol boats waiting offshore. On the same day, fourteen paratroopers led by Amnon Lipkin-Shahak, future chief of staff, attacked the Beirut headquarters of the Popular Front for the Liberation of Palestine, taking on more than ninety armed militants and losing two of their own. Concurrently, other Fatah facilities in Lebanon were attacked by naval commandos.

The high point of the unit's international fame, however, was the July 4, 1976, rescue of one hundred passengers and the twelve Air France crewmembers who had volunteered to remain with them, who were held by seven hijackers and some one hundred Ugandan troops in an old terminal of Entebbe Airport. Within a wider military operation of a total of one hundred Israelis under a brigadier general, the actual rescue in the terminal was executed by twenty-nine members of the unit under its commander, Yonatan Netanyahu. They killed the hijackers and their Ugandan cohorts, with three hostages killed in the cross fire and ten wounded. Netanyahu was the only Israeli combatant killed.

The Entebbe operation was necessarily public and was further publicized by both feature films and documentaries, with Netanyahu himself memorialized by an institute in his name, but the Sayeret Matkal was born and remains a unit for secret operations that are never revealed, let alone publicized. So, it is only because of incidental revelations by other parties that anything at all is known of their operations, such as the 2017 penetration into a headquarters of the Islamic State of Iraq and the Levant (ISIS) that revealed the design of new laptop bombs that could pass airport inspection. Only unconfirmed reports connect the unit to Israel's fairly spectacular operations in Iran.

By contrast the unit's failures are well known, including a 2018 undercover operation in Gaza that triggered wider fighting, before that a weapons-test mishap with fatalities, and going further back the tragic failure of a hostage rescue on Israeli soil. On May 15, 1974, three gunmen of the Democratic Front for the Liberation of Palestine attacked a van, killing two Israeli Arab women while injuring a third, and then entered an apartment building in the town of Ma'alot on the Lebanon border, where they killed a couple and their four-year-old son before entering the local school. Rushed into action, a Sayeret Matkal team attempted an improvised rescue, but twenty-two children were killed and more were injured. That failure resulted in the establishment of a specialized police unit for hostage situations within Israeli territory, the

Yechida Mishtartit Meyuchedet ("Special Police Unit"), known by its Hebrew acronym Yamam, manned by career police professionals.

Against a few failures, there were many successes, and Arnan's creation remains the country's premier commando unit, consistently attracting the best and brightest who are also the fittest. Serving for at least five years but often longer before starting higher education, let alone civilian careers, does not seem to impede subsequent advancement: unit veterans include three prime ministers and more Cabinet ministers, heads of the Mossad and the Shin Bet security service, and famous entrepreneurs, including Daniel Mark Lewin, the mathematician cofounder of Akamai Technologies, killed on September 11, 2001, aboard American Airlines Flight 11, reportedly the day's first victim, stabbed by one of the hijackers probably while trying to resist them.

Moshe Betzer and Unit 5101

Moshe Betzer, better known as Muki, acquired a legendary reputation within the Sayeret Matkal, not least for his jovial serenity under extreme pressure. He rose to its deputy commander before leaving the unit to establish his own Sayeret Unit 5101, Shaldag ("Kingfisher"), after the hunting bird with exceptional eyesight. Following the 1973 war with its heavy losses to Soviet surface-to-air missiles, the air force was urgently pursuing every possible solution to the missile threat, and Betzer had one of his own: a unit that would reach deep behind enemy lines to attack missile batteries, radar installations, and forward headquarters with its own tactical missiles as well as the usual small arms.[30]

Betzer's personal reputation endowed him with greater authority than his lieutenant-colonel rank. He was allowed to take over a company of Sayeret Matkal reservists to start his own special unit on an experimental basis. Like Sharon and Arnan before him, he asked for a lot, took what he got, and waited for opportunities to get more. In combat Betzer had done wonders with ordinary small arms but for his Shaldag he wanted to add new lightweight tactical missiles, laser designators for air force bombs, and whatever else technology could offer—the commando with nerves of steel was also a techie. That was no handicap, but the techie also wanted to become a weapons developer of sorts to equip his unit with a new missile, and that took real money. Salvation came from an inveterate infantryman, Major General Yekutiel "Kuti"

Adam, head of Operations Branch and later deputy chief of the General Staff, who supported Betzer and secured initial funding for his unit and his guided-weapon ideas.[31] In 1976, after two years of hard work, the unit was officially approved by Lieutenant General Mordechai, chief of the General Staff, after he watched a new-style missile drill, but it was still placed under Sayeret Matkal's command.

Betzer's decisive opportunity arrived in February 1977, when the twelve-victory fighter ace Colonel Yiftach Spector, a highly original, first-class mind in both science and literature, became the chief of air operations.[32] A long-time advocate of a dedicated air force commando unit, as soon as Spector heard about Shaldag he wanted to fund it—and the air force had (and has) much more money per capita than the ground forces. But that would mean a formal transfer to the air force, which suited Betzer's operational concept well enough but not his ongoing need for Sayeret Matkal reserve personnel.

Spector did not want a "green" (ground forces) unit under "blue" (air force) command. His very first dialogue with Betzer records an encounter between two radically different mentalities that nevertheless found a compromise:

BETZER: "Instead of a futile struggle with the entire IDF command, let's start with a reserve battalion of Matkal graduates, to train them for air special-operations missions."

SPECTOR: "Do we have any reservists from Matkal available?"

BETZER: "Yes we do."

SPECTOR: "And who will be the founder of the unit?"

BETZER: "I will."

SPECTOR: "Who will lead it?"

BETZER: "I will."

SPECTOR: "I have an issue with Matkal. Their missions are based on long, detailed planning. They strive for high certainty and therefore deal with every minor detail. [But air operations are launched at a moment's notice] . . . I need soldiers who can receive a mission in the morning, plan it by noon and board the choppers to fight that night."

BETZER: "They will do that."

SPECTOR: "I need them always available, under my direct command, they need to wear a blue beret (the IAF beret)."

BETZER: "A blue beret? Not yet. We will do it gradually. First the greens will get used to the idea and in due time they will succumb to it. We will do everything gradually, first we will get the approval from [Deputy Chief of Staff Rafael Eitan] Raful for the unit as such. Then we will house it on an air base."[33]

Shortly after that, Betzer presented his new battalion idea to Eitan, with Spector being the next speaker. When Spector waxed enthusiastic—too much so evidently—about all it could do for the air force, Eitan refused, agreeing to detach it from the Intelligence Corps but only to put it under the Infantry and Paratroopers Corps, whose able chief Brigadier General Uri Simchoni already knew about (and actively supported) the new battalion. Betzer still feared that separation from the Sayeret Matkal would hurt his new battalion, whose personnel were still composed exclusively of Sayeret Matkal reservists, but Eitan assured him that all Sayeret Matkal personnel of all ranks would be assigned to his unit as soon as each left active service.

Eitan kept his promise and added to it by allowing the new battalion to compete for regular-service conscripts as well, thereby transforming Shaldag from a reserve force to an active-duty unit.[34] That put Shaldag in the sharp competition to recruit the best of the brightest from the conscript intake. Again Eitan helped out: after speaking with his son Yoram, a flight instructor, he instructed the air force to send every dismissed pilot-course cadet to be interviewed by Betzer.

Betzer's first training course was similar to Sayeret Matkal's, including in its extralong duration of twenty-two months and its emphasis on autonomous navigation, but the course soon started deviating because the recruits were invited to offer their own ideas about training methods, tactics, and required trainings. As of 2021, Shaldag's training track lasted one year and eight months, while the extra service obligation in addition to the thirty-two months of compulsory service amounted to sixty-four months, reaching in total eight years, including the time during which all members also would go through officers' school and study to complete a university degree. At the end of each exercise soldiers were asked to provide their suggestions; if they related to equipment, the unit's own Weapon and Equipment Development Section, a particular concern of Betzer's, would endeavor to come up with working products that met the need. The developers were civilians on reserve recall, but finding their tasks very exciting, most served for some fifty days instead of the obligatory thirty, with commanders, both officers and senior NCOs, doing as much as eighty days.

Betzer followed Arnan's example in making sure the chief of staff, along with some of his staff officers, would be present when he was ready to demonstrate the unit's latest addition to its capabilities, often obtained by combining some new homemade equipment with tailor-made tactics to accomplish a priority operational task. It was a way to update the General Staff and obtain funding for Shaldag's progress.[35] Though it was called "The Battalion for Air Force Tasks," the unit long remained officially part of the ground forces, until its fifth commander decided that he would elevate Shaldag's status to be the Air Force commando unit, just as Sayeret Matkal was for the ground forces and Shayetet 13 was for the navy. He got his way in the mid-1980s.[36]

As compared to the Navy's Shayetet 13 and the Sayeret Matkal, with its celebrated spectaculars, Shaldag has had very little public exposure, and its veterans have not authored books or articles or granted interviews. Aside from a successful raid on an operational headquarters in the live context of the Israeli counterattack against Hezbollah that started on April 11, 1996, the best-known Shaldag action was its large-scale role in Operation Solomon of May 24, 1999, when hundreds of Shaldag regulars and recalled reservists flew to Addis Ababa to evacuate Jews endangered by the bloody civil war underway.[37] In thirty-six hours, with Shaldag protecting the entire airport, its road access, and flight operations, air shuttles evacuated more than 14,000 Jews from Ethiopia to Israel.[38] Another known operation occurred during the Second Lebanon War of 2006 when some 200 Shaldag and Sayeret Matkal commandos went deep into Hezbollah country to reach Baalbek for Operation Sharp and Smooth on the night of August 1.[39] Their targets were large weapons caches uncovered by Israeli intelligence. Nineteen Hezbollah fighters were killed, with no known Israeli casualties.

While the Baalbek operation remains wrapped in obscurity, the September 6, 2007, destruction of the North Korean–supplied nuclear reactor complex at Al Kibar (aka Dair Alzour) in the Deir ez-Zor region of Syria by Israeli F-16s and F-15s is well documented. What remains obscure is Shaldag's role, because press reports to that effect refer to the use of laser designators by its men to guide air-launched missiles to their targets—an obsolete technique—while others refer to "damage assessment," indeed important to determine if a follow-up strike was called for, but hardly to be performed on the ground given the availability of multispectral overhead photography.[40] In spite of the secrecy, it is clear that Betzer's dream of a different but equally effective commando battalion to complement the Sayeret Matkal has been accomplished. It is remarkable that after so many years each unit still retains the founder's imprint, with the unit of intelligence officer Arnan still focused on clever new ways of collecting intelligence, while the unit of techie Betzer is still the one that uses new weapons in new ways, and even new weapon concepts, not least to carry out operations a long way from Israel.

[14]

The Armored Corps

Discipline and Technological Improvisation

WHILE THE INFANTRY WITH ITS ELITE UNITS AS WELL AS THE AIR force did most of the fighting in between actual wars, it was the armored forces that dominated Israel's wars from the 1956 Sinai Campaign to the 1967 June war, the 1973 October war, and finally the 1982 Lebanon war—the first and so far only time when the IDF fielded more than a thousand battle tanks under one corps-level commander, MG Avigdor Ben-Gal. However, from its establishment in 1948 until after the 1967 war, the Israeli armored forces had to make do with hand-me-downs. Even after it began receiving the latest-model M60A1 Patton tanks purchased from the United States, it could not afford to replace all the older models. So buying scraps or secondhand equipment, refurbishing it and upgrading it, became a way of life.

At the start in 1948, the recent tank battles of the Second World War had just defined armored forces as decisive war-winners. That was bad news for the Israelis: the Egyptian Army invading from the south had some formidable Sherman battle tanks; the Arab Legion advancing westward had Marmon-Herrington Mark IV and Daimler armored cars, modest enough but with two-pounder (40 mm) guns; the Syrians had Renault 35 light tanks with 37 mm guns; and even the ragtag Arab Liberation Army drove from Syria with some Canadian Otter armored cars. Meanwhile, the IDF at first had no proper gun-armed armored vehicles at all, only civilian trucks partially protected with bolted-on steel plates with no weapon turrets, and very few antitank weapons.

That did not dissuade the IDFs most senior officer, Yitzhak Sadeh (born Izaak Landoberg), a decorated veteran of the Imperial Russian Army and former Palmach commander, from establishing the Eighth Armored Brigade. Instead of the three battalions with the one hundred or so battle tanks of a normal brigade, Sadeh's creation only had two battalions: the Eighty-Ninth, mostly equipped with open jeeps fitted with machine guns (commander Moshe Dayan planned to drive fast in lieu of armor); and the highly aspirational Eighty-Second Tank Battalion, which really could not exist without at least some actual tanks. Given the strict Anglo-American arms embargo that intercepted even gamblers' pearl-handled revolvers, there was no possibility of importing any of the thousands of tanks in Europe's war-surplus depots. Instead, a desperate search got underway for any armored vehicle at all.

The first acquisition was a single GMC armored car with a 37 mm gun stolen from a base of the evacuating British Army. It was whole and came with its load of ammunition, unlike the pickings left behind by the rapidly evacuating army: odd armored vehicles of different types, engineless or rusted out or both, with inoperable guns and many missing parts.[1] Nevertheless they too became the unpromising objects of the scavenger hunt underway to equip the Eighty-Second, which managed to find ten very old, very thinly armored French Hotchkiss H-39 tanks with puny 37 mm guns, two British Cromwell medium tanks in good condition—stolen by two intrepid British soldiers who sympathized with the Jews—and a single reconstructed M.4 Sherman.[2] In Italy, no fewer than thirty-two modified Shermans armed with 105 mm howitzers were found abandoned as wrecks and smuggled back to Israel as "tractors," but not much could be done with them. With a meager thirteen tanks, ten of them light, the Eighth Brigade could never achieve the "mailed fist" shock effect of a proper armored force, especially because it was impossible to keep the tanks operating for as long as a day or two before major breakdowns. The crewmen trained in different armies with little shared Hebrew did not help keep the vehicles running.[3]

There were modest successes nonetheless, which induced the IDF to establish another equally nominal brigade, the Seventh Armored. This would acquire great fame after a modest start as a tankless force equipped with M.3 half-tracked armored trucks, a peculiar hybrid with wheels in the front for steering and tracked bogies in the rear. The US army had 53,000 of those curious vehicles built to quickly supply its mechanized forces with something tracked, if only partially, and armored, if only against small-arms fire with no topside protection.

The IDF grew up very quickly amid the extreme pressures of its first war, and by the end of 1948 it had mastered the art of mobile warfare with columns of trucks and jeeps that outmaneuvered slower-moving enemies, and only needed a few armored cars or half-tracks up front to break through roadblocks and barbed-wire entanglements. It was all very rough but worked well enough to drive back the Egyptians all the way to the international border and beyond into the Sinai, to hold the Arab Legion from advancing further, to send the Syrians back to the international border, and to persuade the Iraqi Army to return home. The one thing that was not achieved was the development of a cadre of adequately professional officers and sergeants capable of training armored vehicle crews to operate them safely and keep them running with field maintenance as needed, and to execute well-drilled tactics on command to enable battle coordination on the fly.

Thus, once the war ended in 1949, the IDF's fledgling Armored Corps, amounting to a few officers in a few huts, made a fresh start by trying to train armor crewmen individually, and then as part of their vehicle teams, progressing to three-vehicle platoons, nine-vehicle companies, and then battalions, mostly by using US Army field manuals, whose great advantage was in their ample illustrations, of great help to many who did not know English. Tank gunners were taught by immigrants who had served in artillery units and were familiar with flat-trajectory, high-velocity antitank guns that function just like tank guns, but maneuvers could not really be practiced because there were rarely enough armored vehicles in running order. In addition, some officers visited the august École de Cavalerie of Saumur, France, where cavalry training had given way to armor training, and they learned how to combine infantry in trucks, or better in half-tracks with a number of tanks, so as to form a Sous Groupement Blindée, the French Army's way of arriving at combined infantry-armor task forces.[4]

The young Armored Corps officers were eager to learn, and learn they did, by trial and error, on the two-decade path of professional advancement that led to the accomplished armored forces that won the 1967 war along with the air force, overcoming superior numbers with superior training at all levels. It then gradually emerged that the extreme equipment scarcity of the early years had left a precious legacy: instead of being habituated to passively await the arrival of new tanks and all other equipment in perfect working order, complete with instruction manuals and replacement parts, the IDF Ordnance Corps and armor officers, with their engineers, technicians, and workers, had to do it all by themselves. Once the fighting stopped and it was no longer imperative to send into action anything that looked like an armored vehicle, the Ordnance Corps gradually learned how to repair, refit, and reassemble odd lots of old tanks and other armored vehicles sold off as scrap or surplus to obtain homogeneous lots of functioning vehicles. With all this hard work came a bonus: while much richer armies were stuck with whatever model they had, unchanged and unimproved until the next model arrived perhaps twenty years later, the IDF, limited as it was by persisting arms embargoes and simple poverty, was not constrained by slow manufacturers' timetables, but free to introduce upgrades as soon as they became available. Without improvisational, even creative technical skills there would have been no clamorous armor victory in June 1967, simply because the IDF did not have a single new battle tank in its inventory—all were locally rehabilitated second-hand tanks. The United States produced M48 and M60 Patton tanks in the 1950s and 1960s but refused to sell them to Israel, as the embargo first imposed in 1947 persisted.

While there was much talk of Jewish money and Jewish lobbies, the oil company and State Department Arabist lobby was much stronger: the US embargo on tank sales to Israel continued unchanged even after the 1955 start of large and then larger Soviet tank shipments to the Egyptian, Syrian, and Iraqi armies, which during the 1960s were all supplied with hundreds of Soviet T-54B and T-55 tanks that arrived brand new, with all needed spare parts, together with teams of technicians and instructors to train both combatants and maintainers. But Pattons did reach the IDF nonetheless, because the West German Army was acquiring new Leopard tanks, making its old Patton M48A1s with their gasoline engines and 90 mm guns obsolete. With US agreement, 150 A1s and A2s were to be sold to Israel. But only 40 arrived before the Germans canceled the sale under Arab pressure. Only then did the United States relent, given all the Soviet tanks arriving in the Middle East, and agree to supply the missing 110 M48A1s to the IDF and add another 100 M48A2 tanks. Israel planned to upgrade all 250 tanks drastically, replacing their gasoline engines with safer, more powerful, and more fuel-economical diesels, and by up-gunning them with the 105 mm gun. However, when the June 1967 war started the project had barely begun, so the IDF had only a handful of the upgraded versions and fought with the old ones.[5] But the biggest problem with the IDF Pattons was that they were simply too few: 250 tanks made for an imposing force, but the combined Arab armies had many more tanks.

The British also produced battle tanks; unlike the United States they agreed to sell a few secondhand tanks from the early 1950s and then more, eventually reaching a total of some 250 by the time of the June 1967 war, with more delivered later or acquired from other armies by purchase or capture for a grand total of 660 by 1970.[6] But the Centurion on sale was a World War II design. It came with twenty-pounder 84 mm guns seriously inferior to the then standard 105 mm high-velocity guns, and with Rolls Royce Meteor aviation-type engines that entailed the built-in hazard of gasoline fuel. On the other hand, the Centurion's armor was very good, as was its gun-turret stabilization, which improved accuracy when firing while moving; hence it was worthy of IDF acquisition.

The ordnance officers and tinkerers of the armor depot duly set about replacing the obsolete twenty-pounders with locally produced high-velocity 105 mm guns based on the winning British ROF L7 design that became a global standard, and to modify the turret cupola rings to accommodate a 12.7 mm .50-caliber machine gun (in theory against aircraft, in practice against light vehicles). But the engines could only be replaced with diesels later: diesel engines powerful enough to move fifty-ton tanks were too expensive. Upgrading hundreds of tanks takes time—all the Centurions employed in 1967 and 1973 already had the 105 mm guns, but replacing the engines took until the mid-1970s.

The one thing not added to IDF Centurions was a night-vision device, an omission that could easily have had strategic consequences on the night of October 6, 1973, as Syrian armored units exploited infrared night-vision devices to continue attacking into the Golan Heights at night while the Israelis, dependent on artillery-fired illumination shells and white-light projectors, were at a technical and tactical disadvantage.[7] Assuming that another war would not occur for many years, the IDF had decided not to waste money on the existing infrared night-vision devices, which, though better than nothing, were limited in capability, and to await the upcoming, much better starlight-enhancing technology.

With some 250 Pattons and 293 Centurions operational by the time of the June 1967 war, IDF armor (and thus all ground forces) would have had a very hard time of it against the Egyptian and Syrian armies lavishly equipped with

Soviet tanks. When the war started on June 5, the Egyptian army alone had more than 900 tanks forward deployed in the Sinai, with more in reserve in Egypt. Moreover, by the time the war ended just six days later, the IDF also had to fight the Jordanians, who had 200 Patton M48s (unimproved A1s) and 44 Centurions, as well as the Syrians, who had some 300 Soviet tanks (and some dug-in German WWII Panzer IVs used to shell the Galilee). Even the Iraqis sent an infantry division with 100 tanks to the Jordanian border.

Actually, IDF armor was not so badly outnumbered in 1967, because its hard-won expertise in converting old tanks, even ones with missing or inoperable engines, guns, or turrets, into fully functioning fighting vehicles was most fully applied to the old US-made M4 Sherman tanks whose armor and retrofitted new guns could be just about adequate in combat, with superior training and superior tactics. But its main virtue for the embargoed and poor IDF was that the Sherman had been manufactured so abundantly (49,234 from February 1942 to July 1945) and distributed so widely that no embargo could deny an ample number to the IDF. Some IDF Shermans were actually imported from France in good condition, even with upgraded guns; others were imported from scrapyards all over the world, including the Philippines; but the largest number had been sold off at scrap metal prices when NATO's European armies received their new US military-assistance-program Pattons.

From all those sources, the tank artificers and tinkerers of the Ordnance Corps supplied the IDF with different batches of properly operating tanks, each delivered in numbers large enough to make it possible to maintain the tanks in good working order and train the crews reasonably efficiently. A full-scale standardization of all Shermans would have been impossibly expensive because the IDF had some of every existing type: Sherman M1s with 76 mm guns; Super Sherman M1s with 76 mm guns and improved HVSS (horizontal volute spring system) suspension; Sherman M3s with the original American 75 mm guns; Sherman M4s with a 105 mm howitzer, a few hundred of which had been upgraded in Israel to become Sherman M50s with 75 mm guns taken from French AMX 13 light tanks; Sherman M50 Cummings with diesel engines; and finally Sherman M51s with a shorter version of the French 105 mm (F1) gun.

In the 1967 war the roughly 500 Pattons and Centurions were supplemented by 360 upgraded M-50/M-51 Shermans and 145 older-model Shermans.[8] In 1967 the IDF also employed 160 French AMX-13 tanks. Fast, but very lightly armored, they were originally bought in 1956 because they were the only brand-new tank anyone was willing to sell to Israel. In 1967 they proved incapable of facing the heavier Soviet tanks used by the Egyptians and the IDF disposed of them quickly after that war. Hundreds of the improved Shermans were to fight successfully in the 1973 war against Syrian T55s and Egyptian T62s. The IDF only replaced them all during the 1970s when it finally procured sufficient Centurions and Pattons.

The improvisational mentality really paid off with the production of self-propelled artillery and combat engineering vehicles, which were certainly essential but which the IDF could not possibly afford to buy from new production, if they existed at all. The IDF did design and produce new kinds of armored combat vehicles that relied on the armor and propulsion of the increasingly old yet persistently useful Sherman chassis.[9]

During the 1967 war the IDF was able to enlarge the herd, so to speak, with the capture of 30 of Jordan's total of 44 Centurion tanks and 100 of its 200 M48 Pattons, but the much bigger haul was the Soviet T-54s and T-55s captured from the Egyptian Army. Initially many of these were used without further ado as the Tiran 1 nonmodified T-54s and Tiran 2 nonmodified T-55s. However, by 1973 they were being improved with 105 mm guns identical to those on the Centurions and M60 Pattons, pintle-mounted .30-caliber Browning machine guns, and with additional stowage boxes to change their shape and reduce fratricide incidents. In the late 1970s and 1980s they were upgraded again with reactive armor and computerized fire-control systems. After the IDF decided to stop using them as tanks, a couple of hundred had the turrets removed and hulls drastically modified into Achzarit heavy troop carriers with additional Merkava-technology passive armor.

The 1967 war ended the absolute US arms embargo, though each purchase still had to be approved or denied individually. The IDF duly purchased 150 new M60A1 tanks in 1971, with additional M60s and M60A1s bought later, and all those Pattons were much needed in the 1973 war. But they too underwent local changes, including replacing the original commander's cupola with a local design and other, more minor changes. Hundreds more would be purchased after the 1973 war as the IDF and the Arab armies conducted an arms race. From the 1980s these would all undergo massive upgrades: new, more powerful engines; computerized fire controls; massive armor additions; and new steel tracks—the latter two adapted from the Israeli Merkava's technology. Secondhand Centurions too were procured in large numbers after the 1973 war and similarly upgraded, except for the extra armor, which they could not carry, and in the 1990s they were gradually phased out of service.

Meanwhile, in 1969 the IDF's "Mr. Tank"—MG Israel Tal, who had been the IDF lead in the Israel-UK codevelopment of the Chieftain tank that the British fraudulently refused to deliver—reacted to the British betrayal by persuading his civil as well as military superiors to initiate the development of an Israeli tank, the Merkava ("Chariot"). That had become possible through the accumulated engineering expertise of two decades of work on the conversion of armored vehicles manufactured by others, with increasing metalworking and design skills, as well as in-depth expertise in the production of composite armor. What was unique about the Merkava's development was the absolute design leadership of one man, veteran armor officer Tal, making it very probably the only major weapon system conceptualized, designed, and developed under the control of an individual and not a committee or, as is more common, multiple committees. In the process, the Merkava incorporated drastic choices no committee could have accepted. But Tal's choices were accepted by colleagues and superiors because of all he had done before even proposing that the IDF design and build its own, different battle tank.

Tal's Revolution: Discipline, Gunnery, and Initiative

Israel Tal, like other IDF officers of his generation, started his career at age seventeen as a British Army volunteer during the Second World War. When he returned in 1945, he completed the Haganah's "underground officers" course before the War of Independence started, commanding infantry units before becoming an armor officer, then rising quickly to become chief of the Armored Corps on November 1, 1964.

Two days later, IDF tanks were sent to engage the Syrians in the north. Controlling the high ground of the Golan, the Syrians could fire down on the National Water Carrier works from two overlooking positions, one just north of Tel Dan at the northeast corner of Israel's territory, the other at the Tel Azzaziat high ground that dominated the valley. The Seventh Armored Brigade of the standing forces that trained the conscript intake sent a company of Centurions to silence the Syrian fire. But after firing some 200 rounds they had inflicted no significant damage. Some officers blamed the newly purchased Centurions, but Tal conducted his own after-action analysis and found that the IDF tanks had been positioned too close to each other, used the wrong type of ammunition, and aimed the guns poorly.

Tal personally demonstrated his preferred method during the next clash, during which his own Centurion destroyed two Syrian tanks. In the following months, his crews kept improving their gunnery, hitting Syrian tanks and water-diverting equipment, first at ranges of two kilometers, a decent range for armor operations but far from exceptional for static gunnery. But when the Syrians withdrew to a range of six kilometers—a distance rarely achieved in combat—Tal's crews eventually obtained more hits and continued to do so, following another Syrian withdrawal at Korazim, on August 12, 1965, at the then astonishing range of eleven kilometers with parabolic indirect fire aimed with external telescopes. With that Tal silenced the critics of the Centurion's L7 105 mm gun, earning a reputation as one of the best gunners in the corps he commanded, while also exemplifying battlefront leadership by personally operating one of the guns.

Tal became an armor expert who would achieve a global reputation by starting with the most basic of the basics: already a colonel when appointed head of the Armored Corps, he went to work as a mechanic in the tank repair sheds to become familiar with every aspect of the maintenance and repair of the five different tanks in service.[10] When the Armored Corps expanded after 1956, it suffered a decline in quality, with increasing mechanical breakdowns caused by poor discipline in technical maintenance tasks. That was the enemy Tal set out to vanquish with an all-round discipline campaign that defined minute, inflexible rules for the operation and maintenance of tanks. He connected the Army's informality and easy camaraderie to the lack of technical discipline in his corps. But instead of accepting the lack of discipline as an unchangeable fact of Israeli life, Tal mounted a veritable campaign to introduce in the Armored Corps formal discipline in dress (unknown in the IDF), saluting, and drill. Further, Tal insisted that boots be correctly tied and that officers' uniforms have matching tops and bottoms. In other words, he wanted his armor officers and men to look like not-too-sloppy soldiers, as opposed to the wild-haired partisan bands that still set the tone in the rest of the IDF ground forces.

Tal's ideas were accepted because colleagues, superiors, and, increasingly, subordinates realized that they derived from a deeper philosophical approach.

This started with the calculation that Israel's surroundings were ideal tank country, so that the tank could become the decisive, war-winning IDF arm if it could achieve the required effectiveness.[11] To do that, however, the tank crews and tank maintainers had to differentiate themselves from the disorderly infantry, to turn the Armored Corps into a reliably disciplined force as the precondition of technical competence, especially given the great variety of tank models the IDF had to muster, maintain, and use effectively.

Battle tanks are both complicated and fragile, so that much depends on proper handling by their crews, who must have serious mechanical skills. Hence, in most armies, tank crews are career professionals even if the infantry still relies on conscripts. But the IDF's tank forces had to be conscripts or reservists, so Tal's solution was to institute a high degree of specialization within his corps, and to enforce strict disciplinary control over each operating routine. It meant that uniquely in the IDF, the method was not to study and understand what was needed from the ground up, but rather to follow established working procedures strictly. Because that went directly against the IDF's cult of adaptability and originality, Tal's methods, rigid orders, and restrictions caused tensions inside the army and outside it too, as reservists complained.

But Tal prevailed because everyone could see that technical discipline improved with formal discipline: tank crews repaired defects reliably, instead of improvising hit-or-miss solutions; the tanks were strictly well maintained; and equipment that used to break down frequently was operated successfully by following precise instructions. With such visible success, Tal acquired a cadre of like-minded officers; with their support he introduced more innovations, along with more technical courses, and a strict inspection system for each item of equipment, including a detailed maintenance logbook for every tank. When the reservists were mobilized in 1967, they found tanks waiting for them in unit sheds in perfect running order and ready for action. All lesser equipment, whether personal-, platoon-, or company-level, was also neatly stored and well maintained.

The battles of 1967 showed that Tal's insistence on discipline and technical expertise did not diminish the command initiative and tactical agility of armor officers. Units operating reconditioned Shermans and thinly armored AMX-13 light tanks managed to hold their own against the better Soviet tanks, while the Centurions and Pattons defeated larger enemy forces to advance rapidly on all fronts.[12]

A German Scientist Saves Israeli Tanks: Reactive Armor

Superior long-range gunnery and heavily armored tanks were Tal's answer to enemy antitank weapons. He argued that the accepted response to antitank weapons—attacking them with infantry—was correct for the generally short-range engagements in vegetation-covered and built-up Europe, but would not work in the open desert, where antitank weapons easily outranged infantry weapons and the exposed infantry could not get close enough to the antitank weapons. The solution was to engage the antitank weapons at extremely long ranges so that superior Israeli gunnery and thick armor gave an advantage to the tanks.

The June 1967 fighting vindicated Tal's views. By June 10, with victories on all fronts, it was clear that the IDF's armored forces had exceeded all expectations, precisely because of Tal's insistence on high standards of maintenance and individual skill training.[13] However, though these skills were proven again in the October 1973 war, the enemy had created a new threat—an antitank array several times denser and more advanced technologically than in 1967. Despite initial reverses the IDF again defeated the Arab armies, with the tanks leading the battles, but at a much-increased cost in blood. Against long-range gunnery the enemy pitted long-range guided missiles; against close-in assaults with guns and machine guns the enemy pitted infantry saturated with RPGs and a willingness to spend many lives for each tank destroyed. New, better-protected tanks could not be purchased abroad, and the Merkava was still in development. Even when completed it would take many years to manufacture enough Merkava tanks to replace the older tanks. The IDF had to improve the protection of its existing arsenal with a technological innovation.

West German experimental physicist Manfred Held, a specialist in detonation science, followed pioneers Franz Rudolph Thomanek and Walter Trinks in developing shaped charges for antitank weapons and civil explosives applications, such as oil drilling.[14] Held employed X-ray spectrometry and high-speed flash photography to measure and visualize nanosecond-scale effects, along with his natural intuition for detonation effects. His specialized expertise, new measuring methods, and ability to explain his concepts clearly in hundreds of scientific articles made Held the premier expert on the subject.[15] The idea of explosive reactive armor (ERA) to break up incoming munitions in front of armor shields that could resist splinters was born in 1949 at the Soviet Scientific Research Institute of Steel but was never applied by the Soviet Army, which after brushing it aside in 1944–1945 was unimpressed by the operational value of shaped-charge antitank weapons.

Held visited IDF battlefields between 1967 and 1969 on behalf of the West German government, and returned in 1973 immediately after the October (Yom Kippur) war, examining numerous Arab and Israeli tanks hit by shaped charges. These visits inspired Held's idea of reactive armor and forged his friendships in Israel. His new idea was to encase explosives in a lightweight box of sheet steel that would detonate if impacted by a shaped charge or other explosive, causing the outer steel plate to bolt outwards, counterpushing against the shaped charge and thus increasing the distance the explosive jet, or the penetrator formed by the explosion of the warhead's liner, had to travel to pass through the ERA casing. When installed on a tank, this countering effect meant that by the time the high-explosive jet reached the tank's steel armor it would have exhausted much of its kinetic energy, thereby not penetrating the armor.

Tal personally cemented relations between Held and Israeli industry. Having experienced in the 1973 fighting the devastating vulnerability of plain steel armor to Soviet AT-3 Sagger antitank guided missiles and even antitank rockets such as the RPG-7—present in such very large numbers that their individual range limitations were nullified—Tal embraced Held's invention with enthusiasm, connecting him to Israeli industry to rush his ERA boxes into production. It was also a personal salvation for Tal, then in disrepute for his facile pre-1973 dismissal of the antitank missile threat, which he could now dismiss once again. The result was that the IDF Armored Corps was the first to field explosive reactive armor boxes in large numbers, from the start of 1978.[16]

Tanks could easily accommodate the light, modular ERA boxes, which were individually secured with bolts to fittings that were in turn welded onto critical places on the tank's hull and turret. Moreover, ERA blocks hit and detonated in combat could be replaced individually by field repair crews. While fittings for ERA boxes were installed on the majority of frontline tanks in both conscript and reserve units, the ERA armor modules themselves were kept off in order to conceal their existence.

In the 1982 Lebanon War, Operation Peace for Galilee, the IDF's ERA-fitted tanks were fielded in large numbers for the first time, saving the lives of many an Israeli tank crewman. Syrian tank hunters and Palestinian guerrillas hit some 60 IDF tanks with antitank missiles, some multiple times, over the first four days.[17] Between June 6 and 25, 1982, a total of 203 IDF tanks were hit, 22 percent of the 1,025 tanks that had been sent into combat.[18] But the ERA boxes thwarted most attacks—only 2 of the 60 IDF tanks hit by Syrian commandos at the start were destroyed, while the others continued to operate even after repeated hits.[19] Penetration rates were 50 percent lower; antitank weapons penetrated only 108 of the 203 tanks they hit, and destroyed only 52.[20] During the war, three tanks with ERA boxes were captured by the Syrians and promptly sent to the USSR, which soon reverse-engineered and replicated the technology to produce its own ERA, Kontakt-5.[21] The Soviet army also introduced tandem-warhead antitank rockets to overcome ERA boxes.

[15]

Why the Merkava Is Different

IN THE LATE 1950S BRITAIN BEGAN TO DEVELOP A NEW TANK WITH extraheavy armor and a powerful rifled 120 mm gun: the Chieftain. In the mid-1960s the British offered to supply them to the IDF if the Israelis bought the old Centurions and agreed to help with the development of the next model of Chieftain, providing the benefit of its recent and ongoing combat experience.[1] For example, when the Syrians tried to divert the Banias River that flows into the Jordan, the IDF were able to damage and even destroy the Syrian bulldozers at extreme ranges of up to 5,000 meters (16,400 feet) with the 105 mm guns they fitted on their Centurions. But they discovered that with such a high degree of elevation, the gun would not come back from its recoil, since the weight of the gun combined with gravity was too much for the recoil system. That was duly corrected in the Chieftain.

IDF armor officers were ecstatic at the prospect of finally receiving factory-fresh, powerful new tanks, so they did everything possible to collaborate with the British, who sent two precious Chieftain prototypes, with engineers, to a secret testing base the IDF provided and staffed, while IDF tank engineers joined the development unit in England, comanaged by IDF Armored Corps chief MG Israel Tal, who flew to England each month to help out. The British wanted the IDF's money—payments for the Centurions went into the Chieftain development budget—but mostly it was the IDF's recent combat experience they needed, their own being badly outdated. In addition, the

Israelis helped the British sell their Chieftains to the Iranians, against German competition, with Tal successfully traveling to Iran's armor school for the purpose.

But the British had not in fact decided to sell Chieftains to the IDF. They were debating the matter in secret government councils without informing the Israelis, who continued to share their secrets and to work hard to improve a tank the British eventually decided to sell only to their Arab opponents, with Colonel Muammar Ghadaffy's Libya the immediate prospect and the Jordanian Army actually receiving them.[2] When the British demanded back their prototypes, finally ending the charade in 1969, the reaction of the chief IDF codeveloper was to start work on an indigenous tank, the Merkava (Chariot), whose first version was operational in 1978. In successively improved versions it would become the IDF's only battle tank.

Following the British betrayal, all the more perfidious because of the concurrent British pressure on the US government to deny tank sales to Israel to safeguard the valuable Centurion sales, the IDF faced the arrival of advanced Soviet T-62 tanks in the Arab armies in very large numbers, with nothing new with which to counter them.[3] The United States did offer an ingenious solution: the IDF would receive surplus M48 Pattons of the West German Army, displaced by the new Leopard tank, which would be upgunned with the effective British L7 105mm gun and reengined with US-built AVDS-1790 diesel engines. The conversion work would be performed by Italy's Oto Melara tank assembly line in La Spezia, an inheritance of the US Military Assistance Program of the first NATO years.

Tal was tasked to oversee this four-country undertaking, and to find solutions for the many technical difficulties already becoming apparent. Dispatched to Italy to oversee the secret conversion plant, he discovered that his Italian counterparts both at the firm and at the factory level opposed the production line their government had agreed to establish. (In La Spezia, the Communist Party was a dominant presence, and it too had been enrolled in the post-1967 Soviet anti-Israel campaign; a state-owned firm like Oto Melara could not dismiss employees for refusing to work.) In the midst of this, a first train transport of tanks from Germany was blocked by heavy snow in the Alps, the tank outlines visible under their tarpaulins; a media uproar was followed by a diplomatic scandal.

In 1969 Tal decided to retire from active duty as a result of serious disagreements with IDF Chief of Staff Haim Bar Lev. He returned to active service in 1972 to serve as IDF deputy chief of staff under David Elazar until after the 1973 war, and resigned again in March 1974 because of an argument over strategy with Defense Minister Moshe Dayan. Nevertheless, Tal's leadership in developing the IDF armor persisted because he became head of Mantak (acronym for Minhelet HaTank [the Tank Administration Program in the Ministry of Defense]), Israel's Tank Program Directorate of Israel's Ministry of Defense. After his first resignation, as a civilian reservist and following the debacle with the Chieftain and upgraded M48s, Tal decided it was time for Israel to produce its own tank.[4] Prior to making this decision, Tal had appointed the ministry's financial advisor, Pinkhas Zusman, and ordnance engineer Israel Tilan to assess the financial worthiness and technological achievability of the project. Both advisors concluded that an Israeli tank was feasible and a worthwhile investment. Upon hearing about the project, the US government revoked its refusal to supply M60 Pattons, but Tal was convinced he could do better, receiving the approval of both Dayan and Finance Minister Pinkhas Sapir in August 1970. Tal was authorized to try to establish a tank industry in Israel.

The Chieftain project team members were promptly transferred to work on the Merkava under the sole leadership of Tal, who was answerable only to Dayan, a degree of centralization unimaginable elsewhere. Tal moreover would not need a financial infrastructure of his own but would instead rely on the procurement, legal, and financial offices of the Ministry of Defense.[5] By 1979, Tal's project was an operational tank ready for combat, nine years after it was authorized, roughly half the number of years of the fastest new tank programs of the time elsewhere. Obviously, the key was the project's extreme centralization: Tal stated that a minor component was designed in fifteen minutes before experimentation and production in three weeks, to produce a trial batch of fifty units.[6]

The entire staff consisted of around 150 engineers, some in the ministry's tank development unit (Mantak) and some in a technical design department of the IDF, with Tal involved in every detail, the only one with final say. For example, one of Tal's initiatives was to lengthen the expected running time between failures of the engine and transmission (aka the "power pack") from the US manufacturer's 400 hours—which was in fact only 300 hours in IDF service. Tal plunged into thermodynamics to find out why the engines were not functioning as expected. He had his chief engineer, Shalom Koren, retrieve the oldest engine in the Armored Corps so he could take it apart—this was highly irregular, but Tal explained that the loss of US$300,000 (a large sum in 1970 standards) could potentially save millions of dollars. Eventually Tal uncovered the reasons for the underperformance and suggested thirty revisions to the engine design that increased its life between failures from 300 to 1,000 hours. Having provided all the information to the manufacturer, Tal requested in exchange unlimited access to the blueprints of the engines, which he duly received.

Protection First

The Merkava tank differed from other main battle tanks because Tal's concept of what a tank should be was different. This concept stemmed from both tactical and technical considerations.

As described earlier, he saw tanks as leading the battle throughout, with infantry following in their wake only to mop up. He dissented from the near-universal consensus that tank forces should be held back to cooperate closely with the less protected mechanized infantry and the slower artillery in order to have their protection against antitank weapons. The universal response to the highly proclaimed successes of Egyptian infantry armed with antitank weapons against Israeli tanks during the first days of the 1973 October war had been to reinforce the infantry assigned to protect tanks from enemy infantry and antitank weapons. But Tal was adamant that the impression left by the heavy tank losses of October 6–9, 1973, when the IDF was badly outnumbered, was misleading. Surveys of tanks damaged throughout the war showed that most had been hit not by antitank weapons but by enemy tanks. The problem during the first days was a mistaken application of techniques and tactics, not the choice of tanks as the dominant force. Part of the solution was to improve techniques and tactics against longer-range antitank weapons, and part was to improve the protection of older tanks against infantry antitank weapons, as we have seen.

Tal rejected the notion that tanks should offer a balanced compromise between mobility, protection, and firepower. He argued for the priority of protection, which allows it to move across the battlefield through enemy fire. An insufficiently protected tank, no matter how fast mechanically, cannot move when threatened by enemy fire. So, protection *is* mobility. Furthermore, protection increases firepower because it allows tanks to close range to the enemy, making the weapons more effective.

Tal insisted that the tank consists of two separate systems: the crew and the tank itself. The crew is more important, so protection of the crew should be the first priority of all tank designs. In Tal's design, roughly 75 percent of the weight of the Merkava participates in protecting the tank's crew, while in a conventional tank the proportion might be only 50–55 percent. Every component was designed to further enhance crew protection as well as to perform its designated function, hence the Merkava's unique feature: the engine in front to protect the crew, instead of in the back, as in all other tanks where the engine is "protected" by the crew's compartment in the front. There are some disadvantages to this arrangement: heat and exhaust gases from the engine increase the tank's forward thermal signature and hot air shimmer, interfering with the gunner's aim. It also shifts the tank's center of balance forward, thus shortening its trench-crossing ability (though improving ground traction when climbing steep slopes), and the overall height of the tank must be raised to allow the gun to be depressed sufficiently over the engine. However, the important issue, in Tal's mind, was that the engine adds protection to the crew. It also allows the addition of a rear hull hatch for evacuating the crew and wounded personnel while the tank's bulk hides them from the enemy, unlike rooftop hatches. The disadvantages were to be mitigated in various ways, such as with an improved engine cooling system.

This reconceptualization is what makes the Merkava different: it is not a compromise design but a protection-first design. It is also the only tank designed by tank soldiers based on their own experiences, including the synthesized experience of Tal's exhaustive ballistics research on armor encounters from 1948. As the victor in all its wars, Israel had the additional advantage that, as Tal noted, "the battleground [and its wrecks] remained with us."[7] He had filing cabinets stuffed with reports on those closely studied hits—each incident of penetration carefully photographed, measured, and accompanied by an evaluation and a report explaining exactly what happened and at what range. The unorthodox positioning of the engine derived from Tal's examination of some 500 different tanks damaged in various combat engagements; he found that in only 2 percent of the cases in which the engine was penetrated the tank was immediately immobilized, as opposed to 100 percent once the fighting compartment was breached.

Tal's ideas on the optimal tank were subject to much criticism, but this abated after the 1982 war in Lebanon, when the Merkava was universally recognized as a success: a first-line tank for a much lower cost than any of its counterparts.[8] Operationally, in combat, Tal's protection-first approach was certainly vindicated: it allowed bold tactics and it saved lives. Even when a Merkava was hit beyond recovery by frontal assaults, the crew could exit unharmed from the back. It was also noted that Merkava tanks provided mobile blast shields for the foot infantry against rocket-propelled grenade (RPG) fire.

After the 1982 fighting, many Armored Corps officers wanted to transfer to Merkava units, and some parents even requested the transfer of their children. That caused tensions between Tal and Chief of Staff Rafael Eitan, who wanted to publicly declare that the Merkava was not better protected than the other IDF tanks, while Tal offered to upgrade the other tanks with Merkava technology. The first upgraded M60 Pattons (Magach 7) were ready by the late 1980s. Centurions could not carry the extra weight, so they were gradually phased out. For the Merkava Mk 3, introduced in 1990, a new suit of armor was developed as well as a new concept—a basic tank superstructure on which armor modules were attached. These could be replaced with new modules whenever an improved armor technology was developed. That way the tank's armor could always be upgraded without building a new tank.[9]

From Passive Protection to Active Protection

The Mark 4 was the first Merkava model developed after Tal stepped down as special armor advisor to the minister of defense in 1989, though he remained unofficially involved in the project until his death in 2010. Developed from 1999 and in production from 2004, the Mark 4 was therefore the first "post-Talik" design, and it equipped one tank brigade (the 410th) in the Second Lebanon War of 2006. When Tal designed the Merkava, he envisioned the tank's protection primarily against other tanks (despite contrary impressions, most IDF tanks damaged in October 1973 were hit by tanks), but the Mark 4 was the first design specifically adapted to counter other threats, namely the antitank guided missiles and improvised explosive devices typical of today's irregular battlefields.

During the four decades since the first Merkava entered service, Israel's strategic landscape underwent a dramatic shift. Egypt (in 1979) and Jordan (in 1994) both signed peace accords, and the likelihood of great tank collisions in the Sinai desert, as in 1967 and 1973, had been significantly reduced. Saddam Hussein hoped to replace Egypt as leader of the Arab anti-Israel coalition, but the Iraqi Army was smashed in 1991 and then dissolved in 2003 in wars against the United States. The last time the IDF fought enemy tanks was in 1982, against the Syrians in Operation Peace for Galilee, but after Egypt abandoned the conflict against Israel, Syria could not initiate a war on its own, and since 2011 the Syrian Army has been mired in civil war.

With the threat from state armies declining, the Israeli focus shifted to the remaining threat—Palestinian and Hezbollah terrorist and guerrilla offensives. Hence many in the IDF argued that the internal composition of the IDF ground forces should change from armored and mechanized infantry to light infantry supported by precision weapons. They prevailed, leading to a large reduction in the number of artillery and armored units—the IDF's tank fleet was dramatically reduced in numbers, all tanks older than the Merkava Mk 3 were decommissioned, and there was a debate on stopping production of the Mark 4s.[10]

But Hezbollah and Hamas increased their military capabilities from low-intensity terror and guerrilla warfare to medium-intensity regular warfare. They focused on long-range artillery to attack Israel's civilian rear and well-armed infantry with antitank weapons, ensconced in favorable terrain—mountainous areas in Lebanon and dense urban areas in the Gaza Strip—to prevent IDF ground forces from interrupting their rocket bombardment. This led to a series of medium-intensity wars, which compelled the IDF to employ armored and mechanized forces, exposing the fallacy of the "end of the need for tanks" theory. Merkava Mk 3 and Merkava Mk 4 tanks were employed in the Second Lebanon War (2006), and Operations Cast Lead (2008–2009) and Protective Edge (2014) in Gaza against infantry forces armed with RPGs and a variety of antitank guided missiles (ATGMs), including the latest Russian Kornets, which can hit a tank from more than five kilometers and penetrate more than a meter of armor steel. They also employed massive buried IEDs to detonate beneath tanks advancing toward them, mortars, and other weapons. Destroying Israeli tanks became a psychological symbol for success in battle. The IDF entered the Second Lebanon War under the assumption that it would be a standoff fire war or at most a light infantry war of counterguerrilla operations; to quote one IDF paratrooper battalion commander, "I went into Lebanon [as if] to arrest Palestinian [terrorists] but collided with a regular army."[11] The collision with reality resulted in tactical failures and more casualties.

The bitter experience during the Second Lebanon War awakened the IDF to the fact that its maneuver capabilities were seriously hampered, as Hezbollah squads skillfully used Lebanon's mountainous terrain to conceal their positions and successfully ambush many Israeli tanks. During the thirty-four days of fighting in Lebanon, Hezbollah deployed 600 well-trained antitank specialists. The IDF entered Lebanon with 250 tanks, of which 50 tanks were hit and 22 penetrated, three beyond repair.[12]

Thus, in the Battle of Wadi Saluki, known as the Battle of Wadi al-Hujeir in Lebanon, a column of twenty-four Merkava Mk 4 tanks of the 401st Brigade advanced westward from Tayyiba. When the force descended into a steep ravine, it was attacked from all sides, including from the rear. Hezbollah had prepared an ambush and was launching missiles from hidden positions on the hilltops quite safely, from a few kilometers away. Volleys of Kornet missiles hit the tanks, of which eleven were hit and a few went up in flames. In other actions another six Merkava Mk 4s were hit. Altogether six Merkava Mk 4s were penetrated; seven were disabled, of which five were repaired in the field and two had to be towed, while the other four suffered only superficial damage and continued to fight. The brigade lost a total of twelve soldiers in the war, eight of them crew members of the penetrated tanks. The after-action investigation of tank unit conduct in the war attributed some of the hits to poor tactics, the result of a long neglect of training. Basic skills such as the use of smoke screens or the use of terrain for cover had been degraded by the lack of practice in recent years.[13]

Improved tactics might reduce the number of hits, but not the proportion that penetrated the armor. Although, despite the improvement of antitank warheads, the proportion of tanks penetrated was lower in 2006 than it had been in 1982 (44 percent hit versus 47 percent) and the proportion destroyed was lower still (6 percent versus 23 percent), the result was still deemed unacceptable by the IDF Armored Corps.[14] It was determined to tip the balance back in favor of the tank versus the missile. More passive or reactive armor was not a viable solution given weight restrictions and existing technology, so it needed to come up with a technological breakthrough. The eventual answer would be the active protection system known in English as Trophy (IDF name Me'il Rooakh—"Windbreaker").[15] Its history actually began in the 1973 October war, when more than 50 percent of IDF casualties were tank crews whose tanks were hit by a wide variety of antitank munitions. The trauma of that war spurred Yiftakh, an arms-development unit within the IDF, to start working on an active defense system for armored fighting vehicles (AFV) that would detect incoming shells or warheads and somehow hit them in flight—something never considered before.

Their first prototype was the Sartan ("Crab"), followed by the Akrabut (a type of scorpion). But both projects were finally aborted in 1988 because the new Merkava and Achzarit troop carriers (a heavily armored personnel carrier based on the hull of a T55 tank) were deemed to have sufficient passive defense abilities.[16] The Russians were working on a similar solution and were the first to mount active protection systems on their tanks: the Drozd (first employed in 1981) and Arena (1997) systems. However, these systems had technical problems, so only limited numbers were purchased and employed. The first successful system, Trophy, was developed by Rafael, Israel's premier developer of missiles, based on the prototypes previously developed by Yiftakh.[17]

Rafael overcame two major challenges: the system had to be entirely autonomous, that is, it had to work automatically so long as it was switched on; and it had to avoid collateral damage to friendly forces in or around the tank (a major problem with the Russian systems). In 2005, Rafael introduced the first prototype of Trophy.[18] Development accelerated soon after, due to Hezbollah's successful antitank tactics during the Second Lebanon War.[19] Trophy employs a network of four small radar sensors covering a 360-degree hemisphere around the protected tank. The radars are integrated with the Merkava Mk 4's battle management system, providing instantaneous detection of an incoming missile or projectile fired at the tank. If equipped with a laser detection system, the system can identify the location of the threat prior to the deployment of the enemy missile or projectile (ATGMs such as the Kornet-E use a laser beam to track their targets). The system informs the crew of the location of the firing source, even while the missile is in the air, allowing them to engage and suppress the threat or eliminate it altogether.

Using networkcentric connectivity, the location of the target can also be transferred to other weapon systems and platforms to initiate an immediate counterattack. The kill mechanism of the Trophy system is activated when the enemy missile reaches a specified distance from the tank: multiple explosive projectiles are then aimed at the incoming missile. Mounted on a rotating pedestal, the device turns in the direction of the incoming threat to project a sheath of melted fragments to destroy it. This hard-kill countermeasure is effective against all types of ATGMs, antitank rockets, and also high-explosive antitank projectiles (but not kinetic rounds).

The first tests were carried out in 2009, and the IDF declared Trophy qualified for operational use. Soon after, the IDF's 401st Brigade began fitting its tanks with the system—the same brigade that had suffered heavy losses in the Battle of Wadi Saluki.[20] On March 2011, Trophy was first used operationally, successfully intercepting an RPG-7 aimed at a tank.[21] Later that month, the system proved successful once again by warning the tank crew of an incoming projectile. This time, the system did not intercept the threat because its calculations determined that it would not hit the tank.[22] Following those successes, the 401st Brigade formally adopted the Trophy and retrofitted the system to its tanks.[23] In 2014, the Seventh Armored Brigade started a two-year process of replacing its outdated Mk 2s with Mk 4s that already had the Trophy APS system.[24]

Operation Protective Edge in Gaza (July 8–August 26, 2014) resulted in Trophy's most intensive test to date. Five hundred five Merkava tanks participated in the operation; 66 Merkavas were deployed along the border with defending infantry to block Hamas raids into Israel. Four hundred thirty-nine Merkavas escorted infantry and combat-engineer units entering the Gaza Strip to locate and destroy offensive tunnels leading from inside Gaza into Israel, and to locate and destroy weapon stores and enemy combat units. They provided direct close-range fire support in lieu of the less accurate artillery, covered the infantry as it searched, and sometimes served as carriers for troops, casualties, and supplies.[25]

A key focus of Hamas's efforts was to prove that its men could stop the Merkava, which carried both operational and symbolic significance. Hamas formed specialized antitank units equipped with the best antitank weapons it could find, including Russian Konkurs (AT-5 Spandrel), Fagot (AT-4 Spigot), and Kornet (AT-14 Spriggan) antitank missiles. With them, Hamas adopted a multipronged tactic by engaging IDF tanks at long range with ATGMs while sending in small antitank detachments for close combat with RPG rockets. In addition, Hamas also used IEDs and mines against Merkava formations to draw them into prepared ambushes where all antitank weapons could be brought to bear.

During more than three weeks of high-threat maneuver operations in densely built-up areas, there was not a single hit to a Trophy-defended tank and, reportedly, no false alarms.[26] In one recorded episode, a company commander's tank was targeted by a Kornet antitank missile from a distance of some 3.5 kilometers. "We knew the tank was targeted, but the system worked like a charm. Because of [Trophy], the threat was neutralized. Everything was automatic."[27] Not all Merkavas in Gaza had the Trophy system—the others had to make do with passive armor and good tactics. Although some tank commanders and crewmen were killed by small-arms fire or explosive fragments when outside the tanks, not a single tank commander or crewman was killed or wounded by antitank weapons penetrating the tanks.

As now configured Trophy includes a radar, four radar antennas, a computer system, and an interception system. In action, the system first scans the tank's surroundings and tracks potential threats. When it detects a threat that would hit the platform, it sends a barrage of metal pellets to intercept the incoming threat. The system covers 360 degrees, is fast enough to engage with threats from close range, and is fully functional while on the move.[28] At the time of writing, Trophy is considered the only effective countermeasure against tandem-warhead systems, such as the Russian Kornet and RPG-29 Vampir, and it can simultaneously engage multiple threats arriving from different directions, whether stationary or on the move. Following its success, different Trophy versions are now also being developed for use on medium- and light-armored vehicles.[29]

Another active defense system developed concurrently with Trophy is the Iron Fist (Hetz Dorban, literally "Porcupine Arrow"). It senses incoming threats via an installed radar sensor but also has an optional passive infrared detector. When a threat is imminent, an explosive projectile interceptor is launched toward it. The interceptor explodes very near the threat, destroying or deflecting and destabilizing it but without detonating its warhead. That is possible because Iron Fist uses only the blast effect of the explosive. The interceptor casing is made of combustible materials, so no fragmentation is formed in the explosion, helping minimize collateral damage.[30] It has a modular design to allow installation on a wide range of platforms from light trucks to heavy armored fighting vehicles. The system has been successfully tested against a wide variety of threats including RPGs, ATGMs, tank-fired HEAT (high-explosive antitank) ammunition, and kinetic energy penetrators. Acquisition of the Iron Fist active protection system was approved in June 2009 for installation on the heavy infantry combat vehicle Namer.[31]

In December 2014, it was revealed that a next-generation active defense system was to be developed by combining the Trophy system and Iron Fist technology.[32] Unlike the Trophy's interception method of metal pellets that spread over a wide area, Israel Military Industries' (IMI) interceptor launches an antimissile missile. In June 2016, the US Army chose the Iron Fist Light configuration to protect its light- and medium-armored vehicles, a decision made because of the system's light weight, ability to fire interceptors without shock, and low cost.[33] In June 2018 Rafael announced the US Army had awarded a contract worth nearly US$200 million for Israel's Trophy defense system to shield its M1 Abrams tanks.[34] In June 2021 the British Army also selected Trophy for its entire fleet of 148 Challenger 3 main battle tanks.[35]

Mobility

One of the recurring criticisms of the Merkava was the relative weakness of the engine compared to its weight, so that it was slower overall both in acceleration and in maximum speed. Here too Tal's concept was different—maximum speed when traveling cross-country, especially over the rocky ground typical of most of Israel's expected battlefields, was determined more by crew comfort than by engine power. Driving fast over rocks is a violent experience for a crew, a violence reduced to tolerable levels by the quality of the suspension system. The Merkava's improved spring-based suspension system gave a smoother ride than other tanks on similar terrain, enabling the crew to drive faster. Another important factor in hilly country is the ability to climb steep gradients, an ability partially dependent on ground traction—the forward-placed engine and the steel-track design improved traction so that Merkavas were able to scale ridges and cross terrain inaccessible to other tanks. In one case in Lebanon in 1982 this enabled reaching positions no other tank could and, from there, destroying a Syrian T62 unit.

As successive models of the Merkava were introduced they carried more powerful engines—in part also to compensate for increasing weight as armor protection was thickened. The Mk 1 had a 750-horsepower Continental AVDS 1790 powerpack and the Mk 2 a 900-horsepower version of the same engine; the Mk 3 introduced a new 1,200-horsepower engine with a new locally manufactured transmission, and the Mk 4 a 1,500-horsepower version of that engine.

Firepower

The Merkava Mk 1 and Mk 2 carried a locally produced version of the British 105 mm gun, the Merkava Mk 3 and Mk 4 a locally designed 120 mm gun. All models carried computerized fire-control systems that were constantly upgraded and retrofitted to older models. The Merkava Mk 4 includes a double observation and aiming system that allows the gunner to aim and shoot at specific targets, while the tank's commander can look more broadly at the battlefield.[36] The Merkava Mk 4 is also equipped with more and better periscopes and all-around TV cameras for operations in urban areas.

Beyond these incremental changes, the Merkava Mk 4 introduced fully updated, or more correctly, leading-edge improvements in command, control, and situational awareness of threats both visible and invisible, the result of the application of new artificial intelligence techniques to the processing of data inputs. The overall effect was to enhance greatly the situational awareness of the entire crew, which is provided with continuing imagery of the terrain, the location of friendly forces, the location of known enemy forces, and data on individual targets, the latter of which can be allocated quickly and efficiently between tanks while minimizing friendly fire incidents.[37] By 2011 the Tsayad 600 (literally "Hunter," but also the Hebrew acronym for digital ground forces) system connected all maneuvering units with the IDF's high command, with a later Tsayad 680 upgrade.[38]

During operation Protective Edge in Gaza in 2014, the data system proved its usefulness. A tank was notified of enemy locations via screen data. Then the tank positioned itself while its barrel locked automatically on the target as directed automatically by the system, ready to fire. Previously only fighter jets or attack helicopters had this type of capability, which was now present in the entire Merkava fleet.[39]

The ability to survive enemy fire enabled Merkava crews to close range to their targets, so as to ensure better discernment between targets and nontargets in a confusing environment of buildings, civilians, and enemy combatants, and to improve accuracy of fire. Between them, the 505 Merkava tanks participating in Operation Protective Edge fired 22,269 rounds, including M339 multipurpose tank rounds. This leads us to another aspect of the tank's firepower, the ammunition, beginning with the indigenously developed and gradually improved Hetz ("Arrow") kinetic rounds for piercing armor and locally produced dual-purpose hollow-charge explosive rounds (Halul for

105 mm guns and Halulan for 120 mm guns) that combined antitank and antipersonnel functions. This continued with improved antipersonnel rounds in a series incongruously named after local flowers: Rakefet, Kalanit, and Hatzav. Whereas Hetz, Halul, and Halulan were merely local derivatives of existing munitions, the latter were new concepts in response to the improvement of portable antitank weapons in accuracy, range, and penetration power and the shifting of the majority of battles from open to built-up terrain.[40]

Starting in the 1980s the local producer Israel Military Industries began to develop Rakefet (APAM-MP-T M117/1 Cartridge), a 105 mm shell designed specifically to target antitank squads hiding behind cover. The shell is aimed to fly over the cover and then to eject six subcharges that explode over the target. The Second Lebanon War in 2006 served as the shell's baptism of fire, in which it proved effective against a host of targets. The Kalanit shell (APAM-MP-T, M329 Cartridge) was deployed in 2009.[41] It introduced one major change, ultimately derived from the country's generic immersion in computerized techniques for every prosaic aspect of life, including the total health database available for every individual resident from birth (something unavailable in far richer and otherwise advanced countries). In this case, the Kalanit can digitally acquire updated target data while already loaded in the chamber.

This was an innovation built on a lesson learned during the Second Lebanon War and Operation Cast Lead in Gaza in 2008–2009 to solve the biggest problem for armored vehicles, which is to fire accurately and rapidly on the move.[42] The manual insertion of a target range into the shell's computer meant that maneuvering with a loaded chamber was nearly impossible. With the new communication system built into the shell, the tank can drive with the shell loaded and respond more rapidly to the sudden appearance of a target.[43]

Even after the Second Lebanon War, the Israeli Armored Corps arsenal was still primarily designed for battling enemy tanks that were nowhere to be seen. Many tanks were still loaded with the Halulan.[44] Because shaped-charge warheads funnel most of their blast into a narrow jet to penetrate armor, their antipersonnel effect is limited compared to other explosive shells of similar size. When the target is inside a building, these shells explode on the outer wall, so most of the antipersonnel blast and fragmentation is wasted outside the building, and the effect on the enemy indoors is minimized considerably.[45]

But in 2011, a second innovative shell, the Hatzav (120 mm HE-MP-T, M339 Cartridge), was introduced for urban fighting. It has a tandem warhead in which the first shaped charge penetrates the wall and the second explodes inside. It was during Operation Protective Edge in 2014 that the Hatzav was introduced. The Armored Corps supplied its tank crews with 500 such shells; they shot about 450 of them, and the effects were verified. Major Barak Asraf, head of the Armored Corps Gunnery Department, reported that Hatzav was "three times as lethal as the Kalanit, and half as expensive."[46] Ever since then, the IDF have purchased large quantities of this dual shell.

Heavy Armored Personnel Carriers

Most discussions on the 1973 October war focused on the improved effects of infantry antitank weapons against tanks. But Soviet experts regarded the main threat of those weapons to be their effect on the much-thinner armored personnel carriers (APCs—BTR) and infantry fighting vehicles (IFVs—BMP). A Rafael researcher, Dan Rogal, reached a similar conclusion and suggested that, given the number of personnel that would become casualties if an APC were destroyed and given the need for APCs to travel with or near tanks, a new generation of APCs was needed that carried armor thicker than that of a tank. The 1982 war convinced the IDF that he was correct, and the IDF began building heavily armored APCs. Most APCs have their engine up front and hatches in the rear, which suggested exploiting Merkava hulls for the new vehicle. But production capacity did not allow for the building of both Merkava tanks and Merkava APCs. The solution was to take older tanks, remove their turrets, use the weight saved to up-armor their hulls with Merkava armor technology, and rework the hull to allow infantry squads to ride in them. The first tanks transformed were Centurions renamed Nagmashot. They were successively upgraded with the new armor used in later-model Merkava tanks. Later, when the captured T55s were decommissioned, they were transformed too, creating the Achzarit. However, these improvisations each had various limitations, especially a limited ability to constantly upgrade with heavier armor to match upgraded antitank weapons, so the preferred solution remained the Merkava APC (Hebrew acronym Namer). A major tactical advantage of the Namer is that it allows infantry to travel quickly and relatively safely through lethal areas of the battlefield in which previously only Merkava tanks could operate, thereby allowing combined arms teams to fight together more effectively.

The final decision to actually build Namers was made following the 2006 Second Lebanon War, and the first entered service in 2009. Budgetary considerations and debates over the need for such a vehicle when mostly fighting guerrillas and terrorists slowed acquisition until the infamous case of the July 19–23, 2014, battle of Shuja'iyya in the Gaza Strip. Seven IDF soldiers were killed in a single M-113 armored personnel carrier, sparking a public outcry and serving as a terrible example of the consequences of inadequate protection.[47] As part of the after-action inquiry, the Ministry of Defense announced it would increase the production rate and equip every new Namer APC that enters service with such a protective system—even though the Namer was already the most protected infantry combat vehicle in the world.[48]

[16]

Units 8200 and 81

IN 1993 THREE YOUNG ISRAELIS IN THEIR EARLY TWENTIES, GIL Shwed, Marius Nacht, and Shlomo Kramer, all veterans of Unit 8200 of the IDF Intelligence Corps, launched their start-up Check Point Software Technologies Ltd. Its main product, Firewall 1, was a security filter for incoming internet traffic, derived directly from Shwed's task in 8200, which was to secure the unit's own communication network while it was busily at work to penetrate the communications of Israel's adversaries. Having become the worldwide firewall market leader within three years of its start, with a market share of 40 percent, Check Point kept growing to reach a 2021 market value of US$16.2 billion.

Over the years, many other 8200 alumni started other companies. An article listing them in 2013 commented that "the best tech school on earth is Israeli Army Unit 8200," and that was before the subsequent upsurge.[1] According to a 2013 survey, 10 percent of all high-tech industry workers in Israel declared that they had served in Unit 8200.[2] And the largest of today's Israeli high-tech companies—including NICE, Verint, and Comverse as well as Check Point—were all started by the unit's alumni.

But Unit 8200 of the Intelligence Corps, currently the largest provider of intelligence in the IDF—whose function encompasses signals intelligence (SIGINT), decryption, cyberwarfare, and cybersecurity—was not established to educate entrepreneurial technologists.[3] Its fundamental job is to provide advance warning of threats for the government and armed forces of a country with very narrow borders and no geographic depth to absorb attacks, which is moreover dependent on the mobilization of its reserve forces. Israel is therefore exceptionally dependent on advance warnings, as well as on intelligence that can help prevent, and if need be preempt, the threats that have never been lacking from enemies small and large, already present in place or within pistol shot, or a thousand nautical miles away in Iran or even twice that if jihadi threats from South Asia are considered worthy of attention. Only that menacing concatenation can justify the concentration of Israel's best human resources in its intelligence organizations in general, in the IDF Intelligence Corps more particularly, and especially in Unit 8200, whose core functions of signals intelligence in all its forms provide the most definitive information available in a world filled with unreliable information.

Everything started with the peculiar intelligence requirements of the IDF as a reserves-centered system: it had to have advance intelligence to mobilize its forces to defend the country. Over the years, the threats facing Israel have varied, evolved, and morphed into diverse forms from cross-border infiltration to the nuclear programs of hostile countries, such as Iraq, Syria, and Iran. Unit 8200's priorities changed accordingly, but it always also participated in IDF special operations, from initial planning all the way through execution. Some idea of the unit's activities can be gleaned from what has been publicly revealed over the years.

The Nasser-Hussein call, June 1967. The unit intercepted a phone call of great political importance between King Hussein of Jordan and Egyptian president Gamal Abdel Nasser on June 6.[4] The day before, the air forces of both countries had been annihilated. In the call, Nasser suggested to Hussein that both announce that the air strikes of the previous day had been executed not only by Israeli aircraft but also by British and American aircraft. Accordingly, on the morning of June 7, Arab media announced that American and British airplanes had participated in the Israeli air strikes against Egyptian and Jordanian airfields, provoking mass attacks against American and British embassies in several Arab countries, with other foreigners and residual Jews also attacked in some places.

Because 8200 intercepted the call and passed on the information, the US government knew immediately who was responsible and reacted accordingly. It was Israel's defense minister Moshe Dayan who caused the publication of the phone call text, despite the acute displeasure of the Military Intelligence professionals who wanted to continue to listen to Nasser's calls. But Dayan's priority was to ensure that the US government know exactly who had fabricated the damaging accusation.

Operation Entebbe, July 4, 1976. The night before, four C-130 Hercules transports flew from Israel to Uganda to successfully extricate 102 hostages held by German and Palestinian terrorists under the protection of the country's ruler, Idi Amin. In one of the C-130s were twenty Unit 8200 soldiers who knew Arabic, Russian, English, and Swahili, the official language of Uganda. Their contribution to the operation was pivotal because they monitored all air movements as well as Ugandan Army communications on the ground, reducing the inherent vulnerability of a very small force that could easily have been overwhelmed.[5]

Foiling an airliner terror plot in Australia, July 2017. Unit 8200 intercepted communication between ISIS terrorists planning to blow up an Etihad Airways flight traveling from Sydney to Abu Dhabi. The information Israel passed to Australian authorities led to the arrest of two suspects in Sydney, Lebanese-Australian brothers Khaled Khayat, forty-nine, and Mahmoud Khayat, thirty-two, who were charged with "preparing for, or planning, a terrorist attack," subsequently convicted, and sentenced to long prison terms. Australian home affairs minister Peter Dutton said in an interview that Israel was "directly" involved in uncovering the alleged plot and that "the Etihad flight was almost blown out of the sky and would have resulted in hundreds of people losing their lives, so we are very grateful for the assistance Israel provided in that matter."[6]

Stuxnet. The existence of a computer worm that sabotaged the Iranian uranium-235 separation facility at Natanz was exposed in 2010, and it was identified as Stuxnet. The exposure resulted from the leak of US National Security Agency (NSA) documents by Edward Snowden that showed Stuxnet had been developed jointly by Israel and the United States.[7] Because Unit 8200 is the only organization in charge of conducting cyberwarfare, the link between 8200 and Stuxnet was immediate.[8] It targeted the Siemens SCADA Systems, which remotely control and monitor industrial machines, in this case uranium separation centrifuges. Stuxnet sabotaged the Iranian nuclear program by disrupting the software that monitored the centrifuges, causing them to overaccelerate, overheat, and burn. Moreover, Stuxnet disguised its own activity by sending erroneous information to the centrifuge operators that showed them everything was working properly. Stuxnet's sophistication was such that even had the operators noticed the malfunction, Stuxnet would have blocked them from gaining access and correcting the problem.[9]

Flamer. Another cyberattack on Iran attributed to Unit 8200 is related to the malware Flamer, or Flame. It was discovered by the Moscow-based Kaspersky Lab, which admitted that it had failed to fully understand Flamer because its code was a hundred times longer than any malware it had ever encountered.[10] The Budapest University of Technology and Economics referred to it as "the most sophisticated malware we encountered during our practice; arguably, it is the most complex malware ever found."[11]

The Flamer malware had the attributes of a Trojan horse, a malicious worm and a virus. It runs on the Microsoft Windows operating system and enables the attacker to collect information on the infected computer. Flamer accesses information that includes video and audio recording, screen shooting, recording of keyboard activity, and even changes in the computer's settings.[12] The malware was not only gathering information from Iranian computers but also inflicting actual damage; for example, it sabotaged Iranian oil terminals in April 2012, forcing a shutdown for several weeks.[13] The Budapest University of Technology and Economics estimated that by the time Flamer was discovered, it had already been active for five years.[14] Once the malware was revealed, it became clear that its primary zone of interest had been the Middle East.[15] The countries damaged by the malware other than Iran were Syria, Lebanon, Saudi Arabia, and Sudan, as well as the Palestinian Authority. Within a week of its exposure, the malware began to self-destruct, which raised the possibility that it had been even more widely spread than first estimated.[16]

According to the Iranian national computer emergency response team, MAHER, there is a link between Flamer and Stuxnet, as they shared some common attributes unique to the two. MAHER confirmed that Flamer managed to overcome forty-three antivirus programs and had caused a severe loss of information.[17] According to various sources Flamer, just like Stuxnet, was linked to Unit 8200 because it is in charge of offensive cyberoperations.[18]

Originating at a time when the word "virus" only had medical connotations, Unit 8200 is the current expression of a sustained effort that began before there was a State of Israel. Its original designation was Shin-Mem 2, as the linear successor of the communication intelligence side of the prestate Haganah's intelligence organization SHY, acronym for Sherut Yediot ("Information Services"). Originally SHY collected mostly human intelligence, but it acquired some SIGINT capabilities that eventually evolved into Unit 8200.[19] At the insistence of Ephraim Dekel, head of SHY, the first collection antenna was purchased and installed on Ben-Yehuda Street in Tel Aviv.[20] The antenna's target was a nearby British police station. As soon as its signals were collected it was immediately discovered that the British were encrypting their transmissions, and therefore the SIGINT unit also needed a decryption team. In the informal way of those days, it was soon assembled and set to work. After the first antenna was purchased, very quickly the SIGINT Department of SHY was intercepting seventy-four wireless radio stations of the British police, which provided much intelligence for the Haganah in its struggle against the British Mandate, allowing it to safeguard underground training courses and its modest caches of hidden weapons from police raids.[21] The number of SIGINT targets grew because in addition to the British police, SHY was also eavesdropping on anyone considered hostile to the cause, including foreign consuls, journalists, and members of the two parallel (but competing) Jewish paramilitary organizations, the smaller Irgun and the minuscule but dangerous Lehi (also known by the British as the Stern Gang).[22]

In 1948, with the outbreak of war, the decision was made to dismantle the SIGINT Department of SHY, as every penny and anyone who could shoot were needed for the fight. But the founder and head of the department, Mordechai Almog, insisted that SIGINT was crucial for nascent Israeli intelligence, and decided to keep funding it himself with help from other members of the department. In 1952, Isser Be'eri, the first head of the Military Intelligence Directorate of the IDF, discovered that the SIGINT Department had been working without any legal authorization. He decided to officially reestablish the unit under the IDF as Shin-Mem 2, abbreviation for Sherutey Modi'in 2 ("Intelligence Services 2"). Its first headquarters in Jaffa included barracks as well as technical laboratories and offices for the decryption teams. In 1953, the growing unit and its headquarters were moved to Ramat Hasharon, just north of Tel Aviv, where it remains to this day.[23] The unit's personnel were initially composed mainly of Jews from Arab countries, mostly Iraq, who could use their knowledge of Arabic to produce crucial intelligence that immensely assisted Israel's national security.[24]

In its early days, the goals of Shin-Mem 2 were to collect signals intelligence from the communication networks of the Arab armies and to obtain information about weapons and radars via electronic monitoring and decryption. But even when the unit had antennas installed on the borders, there remained a considerable distance between the Israeli-Egyptian border and the Nile Delta. To close the gap and generally increase the unit's coverage, a new branch was started for aerial intelligence, which initially acquired the cheapest platforms available, DC-3 Dakotas as well as surplus antiaircraft balloons. The latter were replaced with new balloons in the 1970s, remaining in use for thirty years until new aerial vehicles arrived.[25] At the time of writing, Unit 8200 has a variety of aircraft, including Gulfstream jets.

During the 1980s and 1990s, the world experienced the cellular revolution, and Unit 8200 became a pioneer in digital signal processing, developing its own signal receivers that improved the monitoring of cellular phones.[26] Current missions range from classical signal-intelligence gathering to the monitoring and tracking of cellular and fixed-line telephone networks, tactical military networks, weapon systems, and radars. As before, the unit is responsible for deciphering what it intercepts, and is therefore caught up in a cryptological arms race as information security and encryption continue to evolve rapidly.

In Israel as elsewhere cyberwarfare is now an ongoing phenomenon alongside naval, aerial, and ground warfare.[27] Accordingly there is increased pressure on the total resources of 8200, especially because Hatzav—the OSINT (open source) Unit of the IDF—has been put under 8200 command. Its material can be had for the asking but its exploitation requires linguistic skills, as well as linguistic situational awareness.[28]

Yet all along Unit 8200 has remained a combat organization, all of whose able-bodied soldiers of both sexes undergo some weapons training before proceeding to other things. Moreover, 8200 has a unit of its own combatants trained to gather and collect intelligence behind enemy lines.[29] Today, it is the largest unit in the IDF.[30] It is estimated that Unit 8200's personnel number some 6,000.[31] Despite its size, it is still small in comparison to its peers, so that even if it as good as its British equivalent GCHQ or the colossal NSA in quality, it cannot match either in quantity.[32] There is much cooperation with the NSA and GCHQ, but evidently not on all matters, only on common threats.[33]

Secrecy does persist. The name of the unit's commander, a brigadier general and one of the most important officers of the IDF, is never made public, and the unit's structure is also a closely kept secret but for the fact that it changes as the situation requires—the IDF's plasticity persists as well. Reportedly there are three centers:

The Intelligence Center is the largest, with collection, analysis, geographic, country, and functional departments and sections. Intelligence Deployment is in charge of the unit's bases and posts, including those on the borders. One 8200 base was revealed by the French newspaper *Le Monde Diplomatique* in 2010, which described the western Negev base with its thirty antennas and satellite dishes for global signal collection.[34] The Technological Center is an integral part of 8200, with labs and workshops: "The more advanced our equipment was, the more capable we were of providing reliable intelligence in real time." To develop its own equipment and modify that of others, 8200 strives to participate in advances of the cellular and other high-tech and cyber industries.[35] The Deciphering Center evolved from the deciphering teams of SHY, and like its prototype is the most secretive part of a secretive organization, with limited access even for other 8200 personnel. Given the proliferation of encryption, the Deciphering Center faces a sheer volume challenge that can only be overcome with highly developed artificial intelligence methods. The soldiers of Unit 8200 have always been subject to strict security clearance procedures, but their access to other IDF secrets was not limited until 1973, and the Singing Intelligence Officer case.

The unhappy protagonist was Amos Levinberg, a young 8200 officer posted on Mount Hermon on the Israeli Syrian border. On October 6, 1973, the first day of the war, Levinberg's post was seized by the Syrian Army, and he was captured with thirteen other soldiers. In captivity, Levinberg fell to the Syrians' manipulation; they told him that the State of Israel had been defeated, that Prime Minister Golda Meir and Defense Minister Moshe Dayan had committed suicide, and that nothing was left of the population. Levinberg, endowed with a phenomenal photographic memory, responded by telling the Syrians everything he knew.[36] He remembered all the code names, maps, office layouts, and even hundreds of license-plate numbers of various IDF officers.[37] When he was returned to Israel in a prisoner-of-war exchange he confessed to everything; massive remedial action became necessary, but he was not prosecuted. Following the incident, compartmentalization began within Unit 8200, as indeed all over the IDF, in place of the earlier naïve confidence in everyone's discretion. Levinberg evidently knew too much, including a lot of information that had nothing to do with his work. Now, as elsewhere, information access matches each person's area of responsibility.

Unit 8200's many electrical, electronic, and software engineers, along with other specialists, obtained their degrees as serving soldiers at civilian academic institutions, but with army funding that entails a service obligation. Academic-track recruits (*Atuda Academit,* literally "Academic Reserve") start at age eighteen like all other recruits but are then sent to study at the IDF's expense.[38] Engineers who do not follow the Atuda track are known in 8200 as Academizators, soldiers who managed to finish their bachelors of science degrees on their own. Unlike other IDF units, 8200 does not accept volunteers; it monitors high school and other students and chooses whom to invite for testing and selection. In so doing, Unit 8200 has priority over all other units in the IDF.

Reportedly, soldiers who speak Farsi will be prioritized to 8200 even if they are also qualified for flight training or the Talpiot program, otherwise the most prestigious tracks in the IDF, with first right of refusal in all matters of manpower selection. But 8200 subjects its preliminary selectees to standardized requirements: an aptitude test and a profile rank of at least fifty-three out of fifty-six, that is, in the top tenth of all recruits. Once they pass the general tests, the few recruits who make the grade are summoned for another exam that tests their technological potential, with especially high scorers assigned to the technological center. Then cognitive and social tests are followed by an interview, and finally the security clearance process. After that, all who pass are warmly welcomed to 8200.

The subsequent training starts in the Intelligence Corps' basic training camp with some basic infantry combat training. Then the different professional courses start, each lasting several months, during which soldiers live under an intense 08:00-to-22:00 timetable and spend most of their days studying. One graduate recounted, "I was recruited into 8200 just after I had finished high school. I can guarantee you that the few months of studying I did in the unit were more intense than anything I had experienced in years in high school."[39] At the end of this course the soldiers are sent to designated positions at the various 8200 bases.

An alternative on-the-job training track is for soldiers who are immediately sent to their destined positions upon finishing basic infantry-type training. These soldiers can be first sent for academic studies or be fully qualified at their operational departments. Studies can include language training, for example in Arabic, Farsi, and other languages of interest. Unit 8200's Flag

Program is also a prearmy course. It trains the unit's network intelligence officers, who monitor and translate messages.

Interested Israeli high school students work hard to decode the latest 8200 priorities to enhance their chances of being selected. Those who know rarer Middle Eastern languages can count on being summoned absent impediments, and the same is true of students with excellent grades, especially in mathematics, physics, or computer science—a powerful incentive to work hard in school. The selection process lasts some months and may end in a refusal, but concurrent candidacies are allowed for the Flight Academy and top-tier special operations units.

The essential ideology of the 8200 is that nothing is impossible. In its culture it resembles a high-energy version of a high-tech start-up company more than a regular army unit. Unit 8200 commanders are well aware of this and do what they can to encourage ambition unrestrained. Inevitably, the 8200 disciplinary environment is lax, with the divide between soldiers and officers not merely paper thin, as in many ordinary IDF units, but almost nonexistent. Former Lieutenant G, who served five years in the intelligence center of 8200, defined the environment as free and accommodating, an absolutely different experience than serving in other units, even in the informal IDF.[40]

Moreover, 8200 officers follow an open-door policy that extends to its commanding officer. Anyone in the unit, including a new recruit convinced that a matter needs higher-level attention, is free to present themselves at the door (or use a comparable communication device) of the more senior officer, up to and including the commanding officers, without having to respect the official chain of command. This policy does not derive from generic liberality but from caution. The 1973 October war started with a colossal intelligence failure, an Israeli Pearl Harbor or Operation Barbarossa, in which the enemy's vast buildup of forces did not go unnoticed but was disastrously misinterpreted. The IDF's intelligence chiefs failed in performing their most critical task: to provide an early warning of war to enable the mobilization of IDF reserve forces. They detected but explained away the strongest possible indicators that the Egyptians and Syrians were preparing for war. This was because they were captive to the theory that no war would be started by the Egyptians because they had no hope of gaining the means to counter Israel's air superiority, because they knew they could not win a war without it, and finally because they would not start a war they could not win. The IDF chiefs overlooked the possibility that a war might be started to activate international reactions that would rescue Egypt from its predicament after the loss of Sinai in 1967, even if the immediate outcome was a military defeat. That the eventual outcome was indeed a defeat for the Egyptians and more so for the Syrians did not diminish the costs and losses of unpreparedness. The intelligence failure of 1973, caused not by inadvertence but by overweening intellectual arrogance, is now half a century past but remains a defining event for Israeli intelligence. Postwar investigations found that a lieutenant colonel in Military Intelligence headquarters had assembled persuasive evidence that a full-scale war was imminent, but his superior disagreed, and he had no access higher up the hierarchy.

The open environment of 8200 was manifest in a series of events that occurred in January 2003. An 8200 officer, "A," refused to provide information about the whereabouts of a target for an operation in the West Bank because of a "conscience objection." Unit 8200 was ordered by the General Staff to collect intelligence about a specific target, and officer A was in charge of the mission. On the day of the operation, the officer was supposed to pass the relevant intelligence to the Research Department of the Military Intelligence Directorate, but he refused to do so because unarmed civilians might be hurt if the target was attacked. Because the Research Division could not get the needed information from officer A, it approached Unit 8200's commander directly, who ordered the officer to personally provide the information. Even after a direct order, officer A refused for twenty-five more minutes, when another officer reached the operations room and provided the necessary intelligence.[41] Eventually the operation was not executed because officer A's assessment of the possible collateral damage was accepted. Despite his disobedience, he was praised for his good judgment, yet he was also reproached for not providing the required intelligence in real time. The episode reveals that the unit really does encourage its officers to use their common sense in everything they do.

To stimulate minds and relieve pressure on those forced to apply themselves for days on end to uncover one small detail out of many, the unit organizes lectures, seminars, and various enrichment programs in a variety of fields, with fun days to break the routine.[42] Such extracurricular activities are for both soldiers and officers, in order to bond everyone. An example for such an activity is a TED Talks project incorporated by the unit's soldiers in which they get to choose their fields of interest, whether linked to their military work or not, and then offer lectures to their colleagues, opening a window for them in other fields of interest. This project has been a big success in 8200, with one young lieutenant even declaring he had implemented an idea that was inspired by a lecture in his military job.[43]

Another unique project of Unit 8200 is SOOT—SIGINT Outside the Box. A recent bottom-up initiative, SOOT is in fact a "hackathon." Some thirty soldiers and officers participate in each one, with the aim of tackling resistant obstacles by open-order brainstorming. Sometimes even administrative issues are raised, like the division of tasks between different departments. For each session the soldiers take a few weeks off to research the issue as deeply as possible; then they gather to look for possible solutions.[44]

A Different Elite: "Seeing Far" and Unit 9900

Nadav Rotenberg was a typical Israeli teenager full of life and energy who loved sports, the outdoors, and spending time with his many friends and beloved girlfriend. When his time came to enlist, he volunteered to serve in Battalion 202 of the Thirty-Fifth Paratroopers Brigade. On November 7, 2011, his unit engaged armed militants on the Gaza border. In the ensuing battle he was killed.[45] Shortly afterwards Nadav's father, Dror, met a number of his colleagues. They were looking for a way to commemorate Nadav. As the men talked, they shared stories about their sons and daughters. One of the fathers talked openly about his two autistic sons, who were struggling. Listening to his story Tal Vardi, a former Mossad officer, decided to do something about it with the IDF.

Autism is a growing phenomenon, or at least its recognition has grown. A study released in April 2018 by the US Centers for Disease Control and Prevention found that one out of fifty-nine children in America was on the autism spectrum, a 15 percent increase from two years prior and a 150 percent increase from fourteen years before. In total, about 3.5 million Americans have an autism spectrum disorder (ASD) by what is, obviously, an expanded definition. Nor is the phenomenon uniquely American; similar percentages are encountered around the world. Nearly 42 percent of people in their early twenties with autism have never worked and those who do earn meager wages.[46] For young autistic Israelis the frustration is compounded because, unlike their peers, they are exempted from military service and thus denied a major part of the formative experience of most young Israelis.

The IDF have a long tradition of taking upon themselves missions that are not purely military for the greater good of society and the nation. But Tal Vardi had to find some valid justification for the heavy costs of enlisting autistic recruits. He found it within his own area of expertise: intelligence.

A central challenge of our era is the increasing gap between the technical capacity of data collection and production and the human capacity to select and process the flood of information to turn it into meaningful knowledge that might actually be useful. That is where some ASD individuals can have a distinct advantage. In ways that are still undiscovered by science, it seems that their brains are wired differently, enabling them to perform certain tasks more effectively. Most notably, they can focus on a single item for a relatively long time, and also perform repetitive tasks that require screening large amounts of information with very small discrepancies, a task that would quickly exhaust others.

Vardi called his friend and colleague Tamir Pardo, then head of Mossad, who organized a meeting with researchers and IDF officers to find out if autistic Israelis of military age might be usefully employed. When Vardi and Pardo originally started to inquire into the matter, they discovered that another Mossad colleague, Leora Sali, a physicist in charge of the technology team for the Mossad, had preceded them. Sali, who has an autistic son, persuaded some IDF officers to put together a small team of researchers to explore how the IDF could use the special capabilities of ASD subjects. Vardi joined hands with Sali but suggested a different course of action; instead of research, they should start a pilot program. In 2012 the Roim Rachok ("We See Far") pilot program duly started.[47]

The team identified intelligence Unit 9900 as the perfect fit. Unit 9900 gathers visual intelligence, including geographical data from satellites and aircraft, and is responsible for mapping and interpreting the visual intelligence for troops on the battlefield as well as for senior commanders. A key 9900 task is to screen vast numbers of photos of the same subject matter in order to detect very small variations between them, such as a small pile of earth that was moved or a new dirt road seemingly leading nowhere. In dense urban settings, the changes are even harder to decipher.

It is not uncommon for ASD subjects to focus obsessively on the same object for hours, and to possess superb memory for minor details. It was therefore hoped that in comparing photos they could detect the smallest variations.[48] In a visit to Unit 9900, when LG Gadi Eisenkot (IDF chief of staff

2015–2019) stopped by the desk of one of the autistic soldiers, the latter showed him with great pride something of significance he had found in an aerial photo. Eisenkot looked at the photo up close; "Where do you see it?" he asked. "There, it's so clear," the soldier said to his chief of staff, pointing on the computer screen. Eisenkot, no matter how much he tried, could not see anything unusual.[49]

There were many barriers, including how to select those who could adapt to military life. How to expose them to military secrets, knowing they can be more vulnerable to manipulation by outsiders? There were no answers, but Vardi and Sali forged ahead. The program starts with a three-month course at a civilian college, where candidates learn basic social skills and study academic topics that will prepare them for military service. Candidates are on the high-functioning end of the spectrum. Of the roughly 100 applicants each year, some 80 percent are accepted. Many possess specialized knowledge on various topics at a high level, whether it relates to archaeology, languages, or music. After the program, those who qualify enlist in Unit 9900, where they go through special training to qualify as analysts.

Dividends were almost immediate. Before the 2014 Operation Protective Edge in Gaza the ASD soldiers were tasked with comparing tens of thousands of aerial photographs to uncover signs of potential Hamas and Islamic Jihad terrorist activity, such as explosives, mines, or tunnels. Their commanders expected that the work would require a year and a half, but the results were ready in three months, in time for the findings to be used by units in the field.[50] In fact, unlike regular soldiers, the unit members had to be told to stop working after many hours, because they can overwork themselves. While they are working, the ASD soldiers in 9900 typically wear earphones to listen to music, neutralizing anything that might distract them.[51]

The program is now functioning for multiple purposes: it provides these recruits with a sense of belonging and normality by serving in the military as other youths do; it facilitates their entry into the job market; and it can generate very valuable intelligence. Moreover, the skills they learn are directly applicable in many high-tech firms. The Israeli branches of Intel and eBay were the first companies to hire from the program.[52] Other big corporations outside Israel have taken notice. "It's a talent pool that really hasn't been tapped," said Jenny Lay-Flurrie, chief accessibility officer at Microsoft.[53]

As of now, the program has been applied in other parts of the IDF, including the air force. But the IDF had trod this path before. The same IDF chief of staff Rafael Eitan who established the elite Talpiot program was also responsible for starting another continuing program, this time for the less privileged. Officially named MAKAM, the Hebrew acronym for Center for Advancing Population with Special Needs, the program is more widely known by its unofficial name, "The Raful Boys" (from the nickname of Rafael Eitan). It is designed for soldiers who come from difficult backgrounds, many of them school dropouts with criminal records for small offenses, which exempts them—and excludes them—from military service.

After overcoming much resistance in short order, Eitan, one of the toughest officers in the history of the IDF, had his way, and the first recruits enrolled in 1981 for a basic training program of three months designed specifically for them.[54] The training site has symbolic significance: Havat Hashomer ("The Guards Farm"), also known as Sejera Farm, was the site of the first Zionist self-defense organization, Bar-Giora, which was started in 1907 under Ottoman rule. Two years later it grew and became known as Hashomer ("The Guard"), a small, very selective elite organization (its members did not accept even David Ben-Gurion into their ranks, something he never forgot nor forgave). When the need for a countrywide organization became clear in 1920, Hashomer merged into the Haganah, which later became the IDF.

The unit commanders, from squad leaders to the company commander, are all highly skilled and meticulously selected women soldiers—the IDF's first choice for trainers in general. Admission to become a commander is highly selective, and those selected possess a rare combination of toughness with high sensitivity to address the special psychological needs of their soldiers. The commander of the entire program, a lieutenant colonel, is typically an officer with years of experience in combat units such as the paratroopers or the Golani Brigade, and the post is considered quite prestigious. The program itself includes a basic combat training curriculum with additional high-school-level educational content, such as classes in history and geography, as well as individual psychological and emotional support. The young command staff is supported and advised by psychologists and social workers who monitor the entire process.

After four decades of trial and error the program is considered a success. Each year 1,000 to 1,500 soldiers participate, 85 percent of whom successfully complete the course and join the ranks of the IDF in various positions. Many are trained in trades that can provide them with a livelihood after their IDF service: truck drivers, heavy machinery operators, cooks. But each year some

15 percent join combat units.[55] After they complete their conscript service the soldiers of course remain in the reserves for many more years, thereby justifying the program even by the narrowest metric of generating soldiers from potential delinquents. Some advance in rank and even become officers, and their stories inspire other candidates.[56] The larger contribution is of course to Israeli civil society, as every year hundreds of young boys who had a high probability of spending the rest of their life in endless cycles of crime and prison are transformed into law-abiding productive citizens.[57]

Other Kinds of Innovations

Since the outbreak of COVID-19, the IDF, like other armed forces around the world, have been enrolled in a national response. The IDF can do more than most because of the broad base of their reserve forces. While they helped out like other military forces, with field hospitals, for example (though none were actually required by Israel's health care system), their role soon became more central. When the pandemic started to spread rapidly in Israel in March 2020, control measures were quickly enforced, and the pandemic seemed to be under control in May. But during June and July there was a second outbreak, and it became clear that the March measures would not suffice. By August 2020, the magnitude of the task had become clear. The IDF Homefront Command, the civil defense organization, was asked to quickly put together and lead a national command center, the Allon Headquarters.[58] Though headed by the IDF Homefront Command, the Allon HQ coordinates the efforts of civilian organizations, including the Ministry of Health and Ministry of Defense, and local authorities. The 2,000 soldiers manning the Allon Headquarters include those from the Homefront Command, the Medical Corps, the Computer and Communication Directorate, and the Intelligence Directorate; many were reservists who contributed varied civilian expertise.

The task, obviously, was to cut the chain of contagion. For this purpose the Headquarters was organized into four centers: the Testing Center, in charge of management of the national laboratories; the Sampling and Transportation Center, which increases the sampling capabilities of the Ministry of Health and quickly delivers samples to laboratories; the Ella unit for epidemiological investigations, performed by 300 reservists with professional expertise for the task; and the quarantine center, responsible for the management of recovery and quarantine hotels.[59] The Allon HQ uses a common data platform, developed by Unit 8200 and the IDF Computer and Communication Directorate, that integrates data from Israel's four nonprofit health maintenance organizations (Kupat Holim) that enroll all citizens (enrollment in one of the four is legally compulsory) and must accept every applicant regardless of preexisting conditions. Other agencies that collaborate with the Allon HQ include the Ministry of Health and Israel's emergency medical service, Magen David Adom, along with additional agencies and ministries.[60]

Not satisfied with an operational role, the IDF's R&D organization, MaFat of Iron Dome fame, set out to do what it does best—innovate—under its head, Dr. Daniel Gold. MaFat is not lacking in ambition.[61] One idea it pursued is Magen ("Shield"), an app developed by Israel's Internal Security Agency, Shabak staff, and experts from the software company Matrix that correlates location data from mobile phones with data from epidemiological investigations and alerts the user if an infected person is near. It thus helps bring in those who have come into contact with patients for rapid isolation.

Another team developed software called Indor to respond to the problem of corona patients discovered inside a hospital ward, which can cause the shutdown of the entire ward. The new technology provides an advanced monitoring system that detects—via Bluetooth signals, wireless Internet networks, and more—who has just passed by the patient within the ward, thus allowing only those who came near to be quarantined, with the ward continuing to operate normally. Yet another tool is an integration software that reaches out to the databases of the health maintenance organizations and Ministry of Health using artificial intelligence tools to identify early signs of spread to suppress local outbreaks in early stages, provide a clear picture of the viral spread, and give decision-makers a control map of developments in real time.

In parallel with monitoring and warning measures, the teams also worked on the development of various disinfectants. One development is a glove that prevents virus transfer between surfaces and contains an antiseptic. This is in response to the fact that the rubber gloves many people use prevent the contact of bacteria with the body, but do not prevent transfer to the next surface that the person touches.

Another focus is patient identification. Among the developments on the agenda are ways to detect the virus outside the laboratory with the help of electro-optical means or breath tests, and another effort to identify infected people by smell. A main partner in this concentrated effort is Unit 81.[62]

Unit 81

Every few years the IDF unveils the cloak of secrecy around a hitherto secret unit, chiefly to attract candidates in the fierce competition for talented conscripts among Talpiot, the Flight Academy, the Atuda, and 8200. Toward the end of 2020, in the midst of the COVID-19 epidemic, it was the turn of what some consider the little sister of Unit 8200—Unit 81.

While Unit 8200 is a mass producer of innovation, Unit 81 is dedicated to customized innovations, a boutique research and development outfit. Originally known as Branch No. 8 (Anaf 8) and then Unit 432 before it was finally renamed, it is manned by engineers and scientists in different disciplines (many from the Talpiot program) who are considered problem solvers. Their task collectively is to provide customized solutions primarily for IDF Intelligence but also for other commands. Their virtue (their "value proposition," in Silicon Valley–speak) is an ability to deliver quickly. Unlike in other R&D operations that might measure progress in years, the Unit 81 time horizon for an up-and-running solution extends to months at most, sometimes only weeks. Ready to consider any challenge, but preferably Mission Impossible–style challenges, its approach implies the pursuit of effectiveness at all costs, even if it is an inefficient effectiveness only of practical value for one-of-a-kind problems.

One of the unit's founders was the same Avraham Arnan who founded Sayeret Matkal, his purpose being quickly to provide needed ad hoc equipment. Arnan came up with the unit's motto: "Knowledge, desire, and dedication will make the impossible possible." Among its early successes was to add a small invisible camera to the early hand-held Motorola phones used by field agents, decades before every cellphone had one. Another was to attach cameras to remote-controlled toy aircraft before the first drones arrived. Over the years, the unit was awarded the Israel Security Prize thirty-seven times, more than Unit 8200, which is many times larger. In fact, no other unit in the IDF has won more awards.

In recent years, Unit 81's focus had shifted to cyberoperations before shifting again to artificial intelligence applications evolving toward autonomous intelligence machines. Another 81 distinction is the impressive number of start-ups its alumni have founded. "We are the real start-up incubator," they like to boast. A 2021 survey by the Israeli financial paper *Calcalist* showed that over the years 2003–2010, some one hundred Unit 81 soldiers and officers launched fifty start-ups that attracted some US$4 billion in funding, and had a market capital value of US$10 billion.[63]

Unit 81 does not follow any military protocols, its soldiers are rarely in uniform, and its commanders are called by their first names. The lucrative opportunities outside it do pose a severe retention challenge for the IDF, especially for units such as 8200 and more critically for the far smaller 81. But it seems that its exciting and compelling challenges, and the personal satisfaction of meeting them, keep enough of the talented in uniform.

Conclusion

THE SIGHT OF WILD-HAIRED YOUNGSTERS IN RAGGED UNIFORMS appalled the two US Marine officers who were to contact their Israeli counterparts in Beirut in September 1982.[1] It was their task to arrange a formal meeting with the Israeli commanding officer to ensure that the respective patrols stayed well away from one another, to avoid friendly fire incidents of course, but also because the Marines were not there to help the Israelis but rather to accelerate their departure, and all observers needed to know that, and indeed to see it with their own eyes.[2] The Marines did find a junior officer to take them to the colonel they were supposed to meet: with faded shoulder markings standing in for his lieutenant epaulettes, he was sitting on the ground, back against a wall and just as wild haired and unshaven as the soldiers around him; he was eating a meat ration right out of the can, the can opener serving as his fork.

Taught that inner discipline begins with outward appearance, the neat and trim Marines promptly lost all respect for the Israeli Army, though it had just accomplished a 150-kilometer advance very quickly, with a mere handful of casualties.[3] But what happened next might have changed their minds: when the scruffy officer reacted to the sound of an explosion by dropping the can to grip his weapon as he scanned the scene, his even-scruffier soldiers instantly moved with him, forming a well-practiced array to observe all round, weapons in hand ready to fire.

That is the most readily observable difference between the Israeli Army and its counterparts around the world. It is decidedly more informal in appearance and conduct, and not only because of individual preferences. Its officers are never seen in formal dress uniforms for the simple reason that none are issued except to a handful of military attaches in foreign capitals, an airport drill team, and the chief of staff for official encounters with foreign visitors. As for proper saluting and parade-ground drills, training schedules simply allow no time for them. The other difference is that the IDF have been persistently and remarkably innovative in many different ways, not merely incrementally by way of updated and improved versions of what already existed, but rather with a sequence of genuinely new macroinnovations—some organizational, as with the IDF, still the only single-service armed forces; some operational, as with the 1982 air offensive, the first computer-directed military action anywhere; and some technological, as with the Gabriel anti-ship missile developed in the early days of extreme scarcity and the much more recent invisible armor of active defenses for tanks.

Are informality, ragged uniforms, and macroinnovation connected? Are they facets of the same deeper phenomenon? One answer is suggested by an anecdote recounted by an IDF reserve officer:

> In late 1982, my reserve battalion . . . was reinforced . . . with a company of new recruits—immigrants from a variety of countries who had arrived over the age limit for the full term of compulsory military service. They had undergone only a few weeks of basic training before being assigned to a reserve unit.
>
> Quite a few had already served in the armies of the countries they came from. I began asking them questions about their military experiences. They all agreed that none of them had experienced an army as messy and disorganized as the IDF. As a joke I responded: "Well that's why we win wars—war is a disorganized mess, and we learn to cope with it all the time."
>
> We all laughed, but one of my soldiers who had immigrated to Israel from Rhodesia, to which he had previously immigrated from South Africa, having served in the armies of both, interrupted the laughter:
>
> "Actually, that isn't funny, you have a point there. I served in the South African army when the war in Namibia began; it took us half a year to adjust from the peacetime regulations and lifestyle to a wartime footing. For example, at first, every time we camped somewhere we

> would build a regular camp, with flags and whitewashed stones marking our location, etc. It took us months to realize that we were actually helping the enemy locate us and to stop this. The same with . . . other army nonsense. . . ."[4]

That, actually, is the fundamental reason why the IDF have been so persistently innovative: not because of a presence, but because of an absence. Born in war by official declaration before they had any defined structure, shaped by successive improvisations ever since, Israel's armed forces are not composed of solidly entrenched military institutions that may be valued for their resilience but are resistant to change. Nor has there ever been an abundance of funding to offer easy remedies for every problem that comes along. The two absences assure nothing in themselves, but they do leave open doors for innovation that might otherwise be closed—necessity, the proverb tells us, is the mother of invention, but there are hindrances too. As a result, as we have seen, when individuals approach even remotely relevant IDF officers with a new idea, they are much more likely to receive a hearing than would be the case elsewhere, and that extends to individuals with no prior qualifications or claims to expertise.

It is undisputed that war powerfully inspires innovation, but military institutions at war absorbed by their own urgent tasks naturally resist distractions, including proposed innovations that may or may not yield future results, but which would certainly divert attention and resources urgently needed for an ongoing war. In Israel's case, a protracted war has been underway since before its birth, albeit with a diminishing cast of enemies, but with low-intensity fighting that lasts much longer than brief bouts of high-intensity combat. Hence, as with other armed forces at war, the IDF contain many who strive to prevail over the enemy not only by doing their duty but also by coming up with their own personal solutions for the problems they encounter at every turn. These range from boots unsuited to the terrain (boots are *always* unsuited to the terrain) to tactics that might be better than the established ones for the circumstances at hand; new operational methods that make better use of the available forces; new kinds of military units; or new sensors, systems, or weapon configurations that make better use of the technology of the day.

As we know from diaries, war memoirs, and archived technical proposals, such thoughts have occurred to the soldiers, officers, and even chiefs of staff of many armed forces around the world, as well as to many scientists, engineers, and civilian observers endowed with inventive minds.[5] The IDF difference is that they have remained open to the new, so much so that their history since 1948 has been characterized by a series of abrupt innovations rather than by the smooth implementation of well-laid development plans.

In more recent years the IDF have evolved further in the pursuit of innovation, from a readiness to change internally and, more unusually, to readily examine outside proposals, to the establishment of military units whose very purpose is to innovate, that being an innovation in itself. Some are technology focused to begin with, such as Unit 81; others are combat forces first and last but especially attentive to new technological options, as in the case of the commando Unit 5101 Shaldag; one is a program, Talpiot, that exists specifically to harness the talents of exceptional conscripts, while Unit 8200 must innovate constantly to be able to function at all.

In the United States, where the need to promote innovation within the defense establishment has long been recognized, successive high-level organizations have been established for the purpose, headed by especially qualified senior officers with proven credentials as innovators. One such was the Strategic Defense Initiative Organization (SDIO) announced on March 23, 1983, and established in 1984, whose purpose was to develop innovative space-based and other defense systems against nuclear-armed intercontinental ballistic missiles and submarine-launched ballistic missiles. Its aim was to find better alternatives to the ground-based ballistic defense radars and rocket-propelled interceptors that had been developed since 1957 in various configurations (Sentinel, Safeguard), without ever attaining sufficiently satisfactory results. Its first director was Lieutenant General James Alan Abrahamson USAF, well qualified for the task as a past director of the NASA Space Shuttle program. A more recent example is the Joint Artificial Intelligence Center (JAIC) announced in June 2018 as a new subdivision of the United States Armed Forces, whose first director from December 2018 was Lieutenant General John N. T. Shanahan, highly qualified for the task because of his supervision of Project Maven ("an Algorithmic Warfare Cross Functional Team"), which Shanahan defined in 2017 as "a pilot project to serve as pathfinder, that spark that kindles the flame front of artificial intelligence across the rest of the [Defense] Department."[6]

Contemporary estimates of the quality of leadership and the technological instincts of generals James Alan Abrahamson and John N. T. Shanahan were very positive, and later opinion has concurred. Yet those personal qualities could not alter the essentially hierarchical, top-down structures of the organizations they headed; as integral parts of the US Defense Department it could not be otherwise.

IDF units that pursue macroinnovation as the SDIO did and JAIC does are very much smaller, of course, and certainly command only a small fraction of their resources, but they differ more fundamentally because they are mostly staffed by teenaged conscripts supervised by conscripts in their early twenties, with few career officers in their thirties and at most one or two slightly older commanders, who themselves adhere to open-door policies. None of those things, including the inevitable cliché of a start-up atmosphere, ensures creativity, but taken together they remove the most obvious obstacle to innovation, the authority of the old over the new. That was the well-hidden advantage of the IDF's tabula rasa origins back in 1948 when all concerned would have greatly preferred to have solidly established armed forces, even if a tad antiquated. It is an advantage still preserved today.

NOTES

ACKNOWLEDGMENTS

INDEX

Notes

Introduction

1. Tzavah Ha'Haganah le Israel is literally "Army for the Defense of Israel." Zahal is the everyday acronym.
2. The battle of the Seventh Armored Brigade on the Golan Heights, October 6–9, 1973, was explicitly cited as a model in US Army Field Manual FM 100–5, June 14, 1993, Sections 6–20, 6–21, 6–22.
3. The IDF's first remotely piloted air vehicle, the Tadiran Mastiff, first flew in 1973, becoming operational soon thereafter, whereas as late as the 1991 Gulf War the only RPVs in US service were Israeli imports.
4. On October 31, 1968, the targets were the Nag Hammadi bridge over the Nile, a transformer station, and a second bridge at Qena. Elizar Cohen, *Israel's Best Defense* (New York: Crown, 1993), 366–373.
5. The Soviet Union was the first to deploy antiship missiles (the KSShch Shchuka NATO, reported as the SS-N-1 Scrubber) by 1955, followed by the Termit-15 that NATO reported as the Styx, in effect a small jet aircraft. But the Gabriel that became operational in 1969 was a sea-skimmer, much less vulnerable to interception than the Styx.
6. The first active protection system for armored vehicles was the Soviet Drozd, fielded in Afghanistan as the Komplex 1030M-01.
7. For example engineer Manfred Held, German originator of the reactive-armor concept in the West. He received no support in his native Germany in spite of his globally recognized explosives expertise but was immediately taken up by the IDF's armor chief.

1. Raising an Army under Fire

1. Switzerland and Finland, not coincidentally two stalwartly independent countries, also have reserve-centered armed forces, in which the active-duty portion is even smaller than in the IDF because their conscripts serve for less than a year.
2. Jehuda L. Wallach, *El hadegel: Hakamat tsava amami tokh kedei lehima: Hatsava hafederali be'artsot habrit bemilhemet ha'ezrahim vetsahal bemilhemet ha'atsma'ut—mehkar mashveh* (Tel Aviv: Ma'arachot, 1997), 50.
3. National Defense Act (R.S.C., 1985, c N-5), https://laws-lois.justice.gc.ca/eng/acts/n-5/page-3.htm.
4. *Canadian Forces Dress Instructions* (Ottawa: Department of National Defense, 2011), 5-1-2.
5. Still not a separate service, it is now called the "Air and Space Arm," Zroa HaAvir VeHahalal.
6. That is when Aryeh Dvoretzky, a leading mathematician then serving as IDF chief scientist, proposed the use of radio-controlled miniaircraft with stabilized video cameras to photograph Egyptian antiaircraft missile batteries, in place of manned aircraft. Edward Luttwak brought him the idea in June 1970.
7. A Chiefs of Staff Committee, formed in 1923, was a mere talking shop—it lacked a multiservice G-3 to coordinate operations. Its 1936 predecessor, the staffless, budgetless Ministry for the Coordination of Defense, was ineffectual.
8. The Goldwater-Nichols Department of Defense Reorganization Act of October 4, 1986.
9. The songs of the Red Army were fervently sung but its command structure was unknown.
10. As in the Egyptian and Iraqi forces, which later migrated to the Soviet model. Michael J. Eisenstadt and Kenneth M. Pollack, "Armies of Snow and Armies of Sand: The Impact of Soviet Military Doctrine on Arab Militaries," in Emily Goldman and Leslie Eliason, eds., *The Diffusion of Military Technology and Ideas* (Palo Alto, CA: Stanford University Press, 2003).
11. Palmach was established by the British in 1941, equipped and trained to serve as commandos and reconnaissance units for them; after the defeat of Rommel in late 1942 the British tried to dismantle the force. Palmach went underground—masquerading as a volunteer agricultural assistance organization for the Israeli kibbutzim. Training continued secretly as originally taught by the British, though with less resources.
12. Ken Jeffery, *The Secret History of MI6* (New York: Penguin Press, 2010), 689–697.
13. When the leading role of British officers in the fighting became a scandal in London, they publicly withdrew, only to immediately return quietly. *The Spectator,* "The Arab Legion, by One of Its Officers," June 18, 1948, 6.
14. Gerald M. Pops, "Marshall, The Recognition of Israel," http://marshallfoundation.org/library/wp-ntent/uploads/sites/16/2015/01/Israel_Pops.pdf.

15. Unlike his State Department subordinates, Marshall favored Jewish immigration to the United States.
16. The CIA anticipated that Jewish resistance could not last more than two years: "The Consequences of the Partition of Palestine." SECRET 28 November 1, 1947, https://www.jewishvirtuallibrary.org/cia-report-on-the-consequences-of-partition.
17. A greatly underestimated figure, Ehud Avriel, aka Georg Überall, at age thirty was sent to Prague to buy arms, with the support of Foreign Minister Jan Garrigue Masaryk until his March 10, 1948, defenestration. Stalin's lack of enthusiasm for Jews was outweighed by his anti-British priorities, and sales continued after the murder of Masaryk and communist takeover of Czechoslovakia.
18. Reengined Messerschmitt Bf 109s that remained in production—when the original engines run out, they were reengined with ill-matched Jumo 211F replacements that caused landing accidents. Nevertheless, the IDF's first five air combat victories were achieved with those very Avia S-199s.
19. https://blog.nli.org.il/en/hoi_egypt_tel-aviv/.
20. See description and photos of the IDF Armored Corps Museum in Latrun: https://yadlashiryon.com/armored-corps/armored-corps-ever-since/armored-corps-establishment.
21. Moshe Dayan was IDF chief of staff during the 1956 Sinai Campaign and defense minister in the 1967 and 1973 wars; as foreign affairs minister from 1977 to 1979 he directed the secret negotiations that led to Anwar Sadat's visit to Israel.
22. "The Administration for the Development of Weapons and Technological Infrastructure," abbreviated MaFat, a joint body of the civilian Ministry of Defense and the uniformed Israel Defense Forces.

2. How Scarcity Can Force Innovation

1. Heyl Avir, "Air Corps," is now officially Zro'a HaAvir VeHahalal, "Air and Space Arm." A "force," not a separate service, it still relies on IDF-wide support and remains subject to the IDF General Staff.
2. See the revealing account by ex-RAF Derek O'Connor at https://www.historynet.com/spitfire-vs-spitfire-aerial-combat-israels-war-independence.htm.
3. Tolkowsky served in the War of Independence, retired, and rejoined in 1951. He was Air Corps chief in command during the 1956 Sinai Campaign.
4. Jeffrey L. Ethell, *Mustang: A Documentary History of the P-51* (London: Jane's Publishing, 1981).
5. Mikoyan and Gurevich, who exceeded other Soviet designers in utilizing captured German technology, surprised the United States with their MiG-15 (faster than the US F-86 in Korea), then the MiG-17, followed by the radar-equipped MiG 19, and then the globally successful MiG-21.

6. It applied the best German swept-wing technology, with a Tumansky R-11 axial-flow engine derived from the German BMW 003, and remained in service for some sixty years.
7. For example, the Sud-Ouest Aviation Vautour II light bomber; the Heyl Avir reluctantly purchased thirty-one by 1967 for want of anything better.
8. Weizman was a War of Independence air ace, IDF deputy chief of staff in 1966, minister of defense in 1977–1980, and Israel's president, 1996–2000. See Ezer Weizman, *On Eagles' Wings* (New York: Macmillan, 1977).
9. The Atar 09B afterburning turbojet with 13,200 lbf of thrust, another derivative of the German BMW 003. The DEFA 30 mm cannon were likewise derived from the German Mauser.
10. Ze'ev Lakhish and Meir Amitai, ed., *Asor lo shaket: Prakim betoldot heyl ha'avir bashanim 1956–1967* (Tel Aviv: Ministry of Defense, 1995).
11. Dassault never replicated the global success of the Mirage III.
12. The F-35 helmet-mounted display subsystem, developed in Israel and produced by a joint venture.

3. A Youthful Officer Corps

1. Shabtai Tevet, *Moshe Dayan: The Soldier, the Man, the Legend* (Tel Aviv: Schocken, 1971), 415.
2. Martin Van Creveld, *Moshe Dayan* (London: Weidenfeld & Nicholson, 2004), 97.
3. Amiram Bareket, "Batsava mitkonenim: Anshei keva yetsu lepensiya begil 42?" *Globes,* May 19, 2015, http://www.globes.co.il/news/article.aspx?did=1001037877.
4. Edward N. Luttwak, "Reinventing Innovation: Simple Theories, Complicated Remedies," unpublished manuscript.
5. Speech by Douglas Haig before the annual meeting of the Royal College of Veterinary Surgeons, June 4, 1925, https://quoteinvestigator.com/2012/11/30/horse-in-war.
6. 10 USC 526: the limits are currently set at 652 flag-rank generals / admirals, with sublimits of 20 four-star, 68 three-star, and 144 two-star officers, with the balance being brigadier generals or rear admirals; but additional flag-rank officers are allowed for multiservice joint commands.
7. The navy of landlocked (since 1879) Bolivia has several four-star admirals.
8. HaKirya also houses the civilian Ministry of Defense, just as the civilian-led Department of Defense is colocated in the Pentagon with the chiefs and headquarters of each service.
9. Under the US system, all forces are headed by the theater commanders for the Indo-Pacific, Middle East (Central Command), Europe, and Latin America (Southern Command), while the chairman of the Joint Chiefs of Staff is chief advisor to the secretary of defense, who advises the commander in chief, the president. His Israeli counterpart is not the prime minister but the cabinet of ministers as a whole.

10. Israeli strategists define three rings of threat: First Ring—hostile states with a common border with Israel (example: Syria); Second Ring—hostile states with one other state between them and Israel (example: Iraq); Third Ring—all hostile states beyond the Second Ring states (example: Iran).
11. Edward Luttwak recalls that the intended informality was undone by the British Army's insistence on spit-and-polish shiny boots, "brasso-ed" metal detailing, and the "blanco" whitening of belts.
12. The much smaller Dutch armed forces (roughly 41,000 active and 6,000 reserve) had one four-star and eight three-star generals and admirals in 2021, with another sixty or so two-star and one-star officers. Professor Colonel Ret. Frans Osinga, Dutch Military Academy, private correspondence, April 27, 2021.
13. Foreign armed forces that lack a strong NCO corps are routinely downgraded in US capability estimates, mistakenly in the case of armed forces that train their junior officers as combat leaders.
14. Hence the daily incidents between IDF soldiers and Palestinian civilians that feature in tendentiously edited video clips (stone throwers summon the TV cameras beforehand).
15. Todd South, "Extended Training Here to Stay for Infantry and Armor Soldiers," *Armytimes.com,* October 15, 2020.
16. Daniel Kahneman, "The Sveriges Riksbank Prize in Economic Sciences in Memory of Alfred Nobel 2002: Biographical," http://www.nobelprize.org/nobel_prizes/economic-sciences/laureates/2002/kahneman-bio.html.
17. Roughly equivalent to US$200 per month, with twice that for combat soldiers. Immigrant soldiers with no parents in-country receive payments for board and lodging when on leave. Lt. Col. (Ret.) Dori Pinkas, former chief instructor, interview in *Ba'had Ehad,* Tel Aviv, March 15, 2016.
18. OCS Branch Descriptions, Fort Benning Maneuver Center of Excellence, January 9, 2018.
19. Kenneth R. Tatum and Assoc., "Leadership and Ethics across the Continuum of Learning," *Air and Space Power Journal,* Winter 2019, 43.
20. Unless they started flight school after first serving in other services, branches, or units: see IAF, http://www.iaf.org.il/4428-45785-he/IAF.aspx.

4. Innovation from Below

1. See the verdict of the preeminent military historian Martin Van Creveld: "Historically speaking, those armies have been most successful which did not turn their troops into automatons, did not attempt to control everything from the top, and allowed subordinate commanders considerable latitude." Martin Van Creveld, *Command in War* (Cambridge, MA: Harvard University Press, 1985), 273.

2. On mission orders and the culture of the initiative see Eitan Shamir, *Transforming Command: The Pursuit of Mission Command in the US, British and Israeli Armies* (Palo Alto, CA: Stanford University Press, 2011).
3. In less advanced armed forces officers are themselves disinclined to take the initiative, fearing the career risks of failure. For example, in the March 2018 Turkish intervention in Afrin (Operation Zeytin Dalı Harekâtı) the advancing units had overwhelming artillery support but visibly acted in rigid set-piece moves under top-down control.
4. See Haganah Museum website: http://www.irgon-haagana.co.il/info/hi_show.aspx?id=21814.
5. Headed by retired judge Eliyahu Winograd, hence called "The Winograd Commission": see https://online.wsj.com/public/resources/documents/winogradreport-04302007.pdf.
6. As noted by Martin Van Creveld, "Israel's War with Hezbollah Was Not A Failure," *Jewish Daily Forward*, January 30, 2008.
7. See Dan Senor and Saul Singer, *Start-Up Nation: The Story of Israel's Economic Miracle* (New York: McClelland & Stewart, 2009), 67–83.
8. Nir Barkat, "Tsava vehitek," *Ma'ariv*, September 18, 2000.
9. This mentality was undiminished decades later, when Edward Luttwak on a 2019 visit had another hardware suggestion that prompted a conceptual-development contract within days.
10. Moshe Dayan, *Yoman Vietnam* (Tel Aviv: Dvir Co. Ltd, 1977), 62–63. One day before the start of the June 1967 war, Foreign Minister Abba Eban read aloud a message from Secretary of Defense Robert McNamara: "Very much appreciate and personally respect Dayan who provided the most balanced report on Vietnam situation that has ever been brought to my attention." *Moshe Dayan: Avney derekh* (Tel Aviv: Dvir Co. Ltd, 1977), 427.
11. Moshe Dayan, *Breakthrough: A Personal Account of the Egypt-Israel Peace Negotiations* (London: Weidenfeld and Nicolson, 1981), 169–170.
12. Started in 1979 under Chief of Staff Rafael Eitan (1978–1983) at the suggestion of two physics professors at the Hebrew University, Felix Dothan and Shaul Yatziv. "Talpiot," or "turrets," is from Song of Songs 4:4 describing the majesty of a castle's turrets, the height of achievement.
13. These units are secretive, but the IDF provide basic information on their website.
14. The principle is often attributed to Prussian Field Marshal von Moltke. When one of his officers excused an error by pleading he was only following orders, von Moltke replied: "His Majesty made you an officer because he believed you would know when not to follow orders." Trevor N. Dupuy, *A Genius for War: The German Army and General Staff, 1807–1945* (Englewood Cliffs, NJ: Prentice-Hall, 1977), 116.
15. Richard A. Gabriel, *Operation Peace for Galilee: The Israeli-PLO War in Lebanon* (New York: Hill & Wang, 1984), 102.
16. Doron Avital, *Logika bife'ula* (Or Yehuda, Israel: Kinnert, Zmora Bitan, Dvir Publishing House, 2012), 56–58. Avital also told the story in detail to the authors in a private meeting in 2016.

17. General Staff Reconnaissance Unit, or Unit 269, originally established for intelligence gathering behind enemy lines. Avital retired as a lieutenant colonel, completed a philosophy PhD at Columbia University, became a partner in a venture-capital firm, and was elected a member of the Knesset, Israel's parliament.

5. A Reserve Army of Innovators

1. See www.archives.mod.gov.il/pages/exhibitions/bengurion/bigImages/hakamat_tzahal.jp.
2. Dov Tamari, *Ha'uma hehamusha: Aliyata vesh'ki'ata shel tofa'at hamiluim beyisrael* (Moshav Ben Shemen, Israel: Modan Publishing House and Ma'arachot, 2012), 59–214.
3. Author Edward Luttwak tried to use it, but its powerful spring was almost impossibly hard to cock.
4. UPI Archive, Arab Nations Attack Israel, May 15, 1948, https://www.upi.com/Archives/1948/05/15/Arab-nations-Attack-Israel/6118818754330/.
5. Even after recent cuts. Only the Finnish armed forces are comparable: as of 2016 they consisted of some 24,000 on active duty and 230,000 in equipped reserve formations.
6. Maram, later Mamram, an acronym for Center of Computing and Information Systems (*Merkaz Mahshevim UMa'arahot Media),* which provides all IDF data processing. Alexander Speiser, *Mamram—mador hafala,* Association for the Commemoration of the Fallen Soldiers of the IDF Signal Corps (Israel: Yehud Monson, 2020), 60–85.

6. A Different Military-Industrial Complex

1. Including the privately owned Elbit Systems, whose proprietary technology includes the F-35's chief advance—its helmet-mounted display system—and a variety of state-owned entities.
2. As told to Eitan Shamir.
3. In Japan's case the distortion caused by the aircraft-carrier taboo is extreme: its *Izumo*-class vessels are actually fair-sized aircraft carriers at 27,000 tonnes full displacement, but their international categorization is DDH, indicating a destroyer with some air capability, which in Japanese is further reduced to *goei-kan,* which merely means "escort."
4. Almost all big military purchases are "home-state" because the components of major weapon systems are very deliberately purchased from as many different Congressional districts as possible, certainly those of House or Senate members of the Armed Services Committee or the Defense Subcommittee of the Appropriations Committee.

5. See information on MaFat on the Ministry of Defense's website: http://www.mod.gov.il/Departments/Pages/Research_and_Development_Agency_Mafaat.aspx.
6. Uzi Eilam, *Keshet Eilam* (Tel Aviv: Miskal, Yedioth Aharonoth, Chemed Books, 2009), 153–162.
7. Eilam, *Keshet Eilam,* 354–370.
8. In the Patton Museum of Kentucky's Fort Knox, high temple of the armor fraternity, Peled's photo is displayed alongside those of Patton, Erwin Rommel, Creighton Abrams, Georgy Zhukov, and Israel Tal.
9. Eilam, *Keshet Eilam,* 374–378.

7. High-Speed Development

1. Funding the project depended on US assistance. John F. Golan, *Lavi: The United States, Israel, and a Controversial Fighter Jet* (Lincoln: University of Nebraska Press, Potomac Books, 2016).
2. Uzi Eilam, *Keshet Eilam* (Tel Aviv: Miskal, Yedioth Aharonoth, Chemed Books, 2009), 371.
3. Shlomo Erell, *Lefanekha hayam* (Tel Aviv: Ministry of Defense, 1998), 217–218.
4. Ben-Nun (1924–1994) was founder of Israel's frogmen-commandos. On October 22, 1948, he drove an explosive boat that sank the 1,400-ton sloop *El Amir Farouq,* flagship of the Egyptian Navy, jumping off just in time. In 1956 as commander of a destroyer-escort he captured an Egyptian destroyer. Retired in 1966, he fought in 1967 in the Golan Heights as a volunteer. He is commemorated by the Yohai Ben-Nun Foundation for Marine and Freshwater Research.
5. Luciano Garibaldi and Gaspare Di Sclafani, *L'incredibile vicenda di Fiorenzo Capriotti eroe della Decima ed eroe di Israele,* in *Così affondammo la Valiant,* 1st ed. (Turin: Edizioni Lindau, 2010).
6. Mike Eldar, *Shayetet 13: Sipuro shel hakomando hayami* (Tel Aviv: Ma'ariv Book Guild, 1993), 109–162.
7. Avner Shur, Aviram Halevi, and Tal Bashan, *Ha'esh vehademama: Sipuro shel Yohai Bin-Nun, meyased shayetet 13* (Ben Shemen, Israel: Keter Press 2017), 218.
8. Description of the originally wooden-hulled Jaguar boat of the Lürssen shipyard in Bremen: Shlomo Erell, *Lefanekha hayam* (Tel Aviv: Ministry of Defense, 1998), 217.
9. Shur, Halevi, and Bashan, *Ha'esh vehademama,* 221, 225.
10. One was the current chief of German naval intelligence, Otto Kretzmer, winner of the Iron Cross; another was Professor Gabler, a wartime submarine planner. Haim Shahal, the project's chief engineer, believed that a German colleague had been in the SS.
11. Full diplomatic relations, that is. Israel's prime minister, David Ben-Gurion, and Chancellor Konrad Adenauer of the Federal Republic of Germany (FRG) had signed a Reparations Agreement in 1952; under a secret 1960 agreement the FRG agreed to

provide US$60 million worth of weapons, including US$12 million for torpedo boats.

12. Shur, Halevi, and Bashan, *Ha'esh vehademama,* 226.
13. Shur, Halevi, and Bashan, *Ha'esh vehademama,* 226–227.
14. He was candid as to French motives: "M. Maurice Schumann: Notre politique a abouti au regain de notre influence dans le monde arabe," *Le Monde,* January 14, 1970.
15. See Mike Eldar, *Ha'oyev vehayam* (Tel Aviv: Ministry of Defense, 1991), 170–182; Moshe Imbar, *Shayetet 3: Sfinot hatilim beheyl hayam* (Tel Aviv: Ministry of Defense, 2005), 32–33.
16. Shlomo Erell, *Lefanekha hayam* (Tel Aviv: Ministry of Defense, 1998), 312–313; Imbar, *Shayetet 3,* 36–37; Eldar, *Ha'oyev vehayam,* 191–199.
17. Uzi Rubin, "Israel's Air and Missile Defense During the 2014 Gaza War," BESA Center for Strategic Studies, Mideast Security and Policy Studies No. 111, February 2015, 18.
18. Ulrike Putz, "Graveyard Shift for Islamic Jihad: A Visit to a Gaza Rocket Factory," *Spiegel Online International,* January 29, 2008, http://www.spiegel.de/international/world/graveyard-shift-for-islamic-jihad-a-visit-to-a-gaza-rocket-factory-a-531578.html.
19. Theodore A. Postol, "An Explanation of the Evidence of Weaknesses in the Iron Dome Defense System," *MIT Technology Review,* July 15, 2014, cited insurance damage claims to depict the system as ineffectual because separated rocket warheads still detonated. That is, Iron Dome could not *nullify* the attack, but it displaced the damage to save lives and property even if separated warheads did explode.
20. Michael Gilmore, Director of Operational Test and Evaluation, US Department of Defense, as cited in Bill Sweetman, "Not Combat Ready," *Aviation Week and Space Technology,* February 15, 2016, 34.
21. Carmel Liberman, "Mefaked heyl ha'avir: Anahnu harishonim lehishtamesh ba-F35 bemivtsa hetkefi," *Bamachane' Journal,* May 22, 2018.
22. From the first, overhead synthetic aperture radars could reveal the outline of any stealthy aircraft below them because their airframes occlude the terrain otherwise detected.
23. See Reuven Pedatzur, "Ma kara lekipat barzel," *Ha'aretz,* August 28, 2013, https://www.haaretz.co.il/opinions/.premium-1.2107807.
24. Avi Kober, "Iron Dome: Has Euphoria Been Justified?," BESA Center Perspectives Paper No. 199, February 25, 2013.
25. BG (Res.) Dr. Danny Gold served in the air force and then in the Weapons Development Department. After developing the Iron Dome system he retired as brigadier general and became head of MaFat.
26. BBC News, *Q&A: Gaza Conflict,* January 18, 2009, http://news.bbc.co.uk/2/hi/middle_east/7818022.stm.

27. Uzi Rubin, "Kosher histagluta shel ma'arekhet habitahon beyisrael leshinuyim mahap'khani'im basviva ha'estrategit: Hahagana ha'aktivit kemikreh bohan" (PhD thesis, Bar-Ilan University, 2018), 161–172.
28. "Mesirut, tsiyonut vekhama halakim mi-Toys R Us: Re'ayon im hatsevet hamovil shel kipat barzel shekol haverav bogrei hatekhniyon, al sod hahatslaha shel haproyekt," *The Expert News,* July 9, 2014, http://tracks.roojoom.com/r/12638#/trek?page=1; also see Ilan Kfir and Danny Dor, *Kipat barzel veha'anashim she'asu et habilti ye'uman* (Or-Yehudah, Israel: Kineret Zmura-Bitan, 2014), 130.
29. "Hahmtzot shel Yisrael," *Ynet,* April 15, 2018, https://www.ynet.co.il/articles/0,7340,L-5227361,00.html#autoplay.
30. BG (Ret.) Dr. Uzi Rubin, interview, Tel Aviv, May 9, 2016. He founded and headed Israel's missile defense organization, *Minhelet Chuma* (1991–1999), overseeing the Arrow antimissile defense system, for which he received the Israel Defense Prize in 1996. All interviews in the book were conducted by Eitan Shamir.
31. The IAF never actually endorsed the system. Uzi Rubin, "Iron Dome Versus Grad Rockets Dress Rehearsal for an All-Out War," BESA Perspectives Papers No. 173, July 3, 2012, https://besacenter.org/perspectives-papers/iron-dome-vs-grad-rocketsa-dress-rehearsal-for-an-all-out-war/.
32. A., Senior Iron Dome developer in MaFat, interview, Tel Aviv, June 15, 2016.
33. Uzi Rubin, interview.
34. Senior developer in MaFat, personal email correspondence with the authors, August 15, 2016.
35. Avigdor Zonnenshain and Shuki Stauber, *Mehakonkord lekipat barzel: Nihul ma'arakhot tekhnologiyot bame'ah ha-21* (Haifa, Israel: The Technion Institute for Research & Development, 2014), 92–94.
36. Zonnenshain and Stauber, *Mehakonkord lekipat barzel,* 89.
37. Zonnenshain and Stauber, *Mehakonkord lekipat barzel,* 95–96.
38. "Mesirut, tsiyonut vekhama halakim," 124, 125.
39. Kfir and Dor, *Kipat barzel,* 126, 143.
40. "Mesirut, tsiyonut vekhama halakim."
41. Zonnenshain and Stauber, *Mehakonkord lekipat barzel,* 97.
42. Zonnenshain and Stauber, *Mehakonkord lekipat barzel,* 97–99.
43. Israel Under Fire, IDF website, https://www.idf.il/en/articles/defense-and-security/israel-under-fire/.
44. Duah mevaker hamedina, *Mukhanut lizman herum—tahalikh kabalat hahahlatot lefitu'ah ul'hitstaydut bema'arakhot lehagana aktivit keneged raketot karka-karka (RKK);* duah shanti 51alef 2009 [Readiness for Emergency, Annual Report No. 51A 2009, The State Comptroller & Ombudsman of Israel], 85, https://www.mevaker.gov.il/he/Reports/Report_335/ReportFiles/fullreport_2.pdf.
45. Duah mevaker hamedina, *Mukhanut lizman herum,* 93.
46. Edward Luttwak knew Burke when he inspired a then innovative think tank best left nameless.

47. There were serious imperfections: Teller's W-47 warhead had an unreliable trigger mechanism, and the A2 missile was running late. But Burke and Raborn were not fired. Instead, they led the corrective efforts, so that by the end of 1961 the A2 version was ready, and the W-47 was successfully modified.
48. Judah Hari Gross, "US Army Receives 1st of 2 Iron Dome Batteries, but Future Unclear," *Times of Israel,* September 30, 2020. During Operation Protective Edge, Iron Dome interceptors executed 735 successful interceptions: Ben Hartman, "50 Days of Israel's Gaza Operation, 'Protective Edge' by the Numbers," *Jerusalem Post,* August 28, 2014.

8. IDF Women as Innovators

1. Norway followed in 2015, but only one-sixth of the age cohort actually enlist. See Colonel Ode Inge Botillen, "Universal Conscription in Norway," Norwegian Armed Forces, Defense Staff Norway, https://www.defmin.fi/files/3825/BOTILLE_2017-06-12_Universal_Conscription_in_Norway.pdf.
2. One example among many is the centerpiece of Béla Vizkelety's "Siege of Eger Castle."
3. The first armor instructor was Racheli Bar-Ziv: see Shaul Nagr, "Mahzor rishon shel madrikhot shiryon," *Shiryon* 37 (March 2011): 62–64.
4. Avishai Katz, *Hayalei shokolad* (Tel Aviv: Carmel Press, 2011).
5. A volunteer trained to serve as a tank gunnery instructor after graduating from a well-reputed US high school and a highly rated US university declared that IDF training techniques were "altogether more effective, to a different dimension." Personal conversation with one of the authors.
6. Yael Luttwak, former tank gunnery instructor, IDF Armored Corps, interview, Tel Aviv, 2016.
7. There are some career NCOs in the IDF with maintenance and administrative responsibilities, but the IDF mostly have to rely on their young conscripts.
8. Aryeh Hashavya, *Tsahal beheylo: Heyl hashiryon* (Tel Aviv: Revivim Press, 1981), 196.
9. Or Heler, "Arayot hayarden: Tsahal yakim g'dud hadash babika," *Hadshot 13,* November 14, 2014.
10. Chen Kutz Bar, "Na lehakir: Elinor Joseph, lohemet arviya betsahal," *NRG Online,* February 6, 2010, https://www.makorrishon.co.il/nrg/online/1/ART2/050/556.html.

9. Military Doctrine and Innovation

1. Acronym for Mifleget poalei eretz Israel, "Party of the Workers of the Land of Israel." It dominated preindependence Jewish politics, policies, and institutions, including the trade unions, cooperatives, and most collective settlements, and was Israel's ruling party from independence until 1977.

2. Achdut Haavoda, "Labor Unity," merged into the Marxist-Zionist, pro-Soviet Mapam in 1948.
3. Also known as ETZEL, acronym for Irgun tzvayi leumi, "National Military Organization."
4. By contrast, when Yasir Arafat was given control of a Palestinian ministate, he refused to disband his own party militia, the other party militias persisted, and no unified army ever emerged. The official security organizations of the Palestinian Authority are all controlled by the Fatah party to this day.
5. Edward N. Luttwak and Daniel Horowitz, *The Israeli Army* (London: Allen Lane, 1975), 74.
6. Eitan Shamir, *Transforming Command: The Pursuit of Mission Command in the US, British and Israeli Armies* (Stanford, CA: Stanford University Press, 2011), 84–85.
7. During Operation Nachshon and the subsequent fighting in Latrun, some twenty-five fighters under age eighteen were killed in action.
8. By 1921, Sadeh was in charge of the emerging Haganah militia in Jerusalem. In 1937, while commander in the British-salaried Jewish Settlement Police, he founded the FO'SH, Plugot Sadeh ("Field Companies"), the first properly trained force in the Haganah. In 1941, he was a founder of the Palmach, heading it until 1945, when he was elevated to Haganah chief of staff. In 1948 he formed an armored IDF brigade with three tanks.
9. There are several rather contradictory biographies of Wingate. See Peter Mead, "Orde Wingate and the Official Historians," *Journal of Contemporary History* 14, no. 1 (Jan. 1979): 55–82. He is commemorated in the Wingate Institute, Israel's National Centre for Physical Education and Sport, and elsewhere in Israel.
10. Moshe Dayan, *Avney derekh: Otobiografiya* (Tel Aviv: Edanim Publishers, 1976), 104.
11. Ezer Weizmann, *Al kanfei nesharim* (Tel Aviv: Ma'ariv, 1975), 101.
12. See Yehuda Slutsky, *Sefer toldot hahagana: Mehagana lema'avak* (Tel Aviv: Ma'arachot, 1959), 2: 230–231.
13. Yigal Sheffy, *Sikat mem-mem: Hamahshava hatsva'it bakursim lek'tsinim bahagana* (Tel Aviv: Ministry of Defense, 1991), 32.
14. Yosef Avidar (1906–1995), born in Russia, joined the Haganah at nineteen, was a founder of the Ayalon Institute (the Haganah's secret ammunition plant), chief of the Quartermaster Corps in 1948–1949, then commander of the Northern Command and then Central Command. Postretirement, he became ambassador to the USSR and later Argentina.
15. Sheffy, *Sikat mem-mem*, 56–57.
16. Anita Shapira, *Yigal Allon: Aviv heldo—biografiya* (Tel Aviv: HaKibbutz HaMeuchad, 2004), 141.
17. Dori Pinkas, "Mekorot Hamtakal Bt'shal" (MA Thesis, Bar-Ilan University, 2006), 43.
18. Zehava Ostfeld, *Tsava nolad* (Tel Aviv: Ministry of Defense, 1994), 560.

19. Mordechai Naor, *Laskov* (Jerusalem: Keter, 1988), 177–178. Haim Laskov was IDF chief of staff from 1958 to 1961.
20. Yitzhak Rabin, *Pinkas sherut* (Tel Aviv: Sifriyat Ma'ariv, 1979), 1: 94.
21. Aryeh J. S. Nusacher, *Sweet Irony: The German Origins of Israel Maneuver Warfare Doctrine* (MA Thesis, Royal Military College, Canada, 1996).
22. Eric Hammel, *Six Days in June: How Israel Won the 1967 Arab-Israeli War* (New York: Scribner, 1992), 24; Haim Bar Lev, "Defusey lohamat shiryon: Beshuley hatimrun," *Ma'arachot* 130 (August 1960): 13–15; and Uri Ben Ari, *"Nua nua! Sof." Hama'avak al derekh hashiryon,* (Tel Aviv: Ma'arachot 1998), 50, 102–104, 114.
23. Martin L. Van Creveld, *The Sword and the Olive: A Critical History of the Israeli Defense Forces* (New York: Public Affairs, 1998), 159.
24. BG Julian Thompson, "Foreword," in Martin L. Van Creveld, *Moshe Dayan* (London: Weidenfeld and Nicolson, 2004), 11.
25. The *Decima Flottiglia Motoscafi Armati Siluranti,* aka Xª MAS, the world's first combat frogman unit.
26. Nevertheless Shayetet 13 kept touch with Fiorenzo Capriotti till his 2009 death. Mike Eldar, *Shayetet 13: Sipuro shel hakomando hayami* (Tel Aviv: Ma'ariv Book Guild, 1993), 138–152. Also see Capriotti's own *Diario di un fascista alla corte di Gerusalemme,* privately published distributed by Alto Mare Blu, https://www.altomareblu.com/diario-di-un-fascista-alla-corte-di-gerusalemme-fiorenzo-capriotti/.
27. Motty Basuk, "Haramatkal Eisenkot metakhnen mahapekha benihul taktsiv tsahal," *The Marker,* April 16, 2015; Editorial, "Kol ma sheratsitem lada'at al tarsh Gideon," *IDF Online,* July 26, 2015.
28. As an example, in 2019 Edward Luttwak visited an IDF training site where he saw how a severe tactical problem was to be solved, a process requiring much courage and skill. He communicated a safer hardware-based modification solution to the escort officer. In short order, the MG in charge of the relevant branch came to his hotel, a conceptual development contract was signed in three days, and the engineering work was underway in months.

10. From Triumph to Failure in the Air, 1967 and 1973

1. As of October 29, 1956: sixteen Gloster Meteors, twenty-two Dassault Ouragans, and sixteen Dassault Mystère IVAs. Yizthak Shtigman, *Me'atsmaut lekadesh: Heyl ha'avir bashanim 1949–1956* (Tel Aviv: The IAF History Branch, 1990), 322.
2. A technique suggested by a flying accident: a trainer had cut high-voltage wires, causing a power cut. Avigdor Shachan, *Kanfei hanitsahon: Letoldot heyl ha'avir vemahal* (Tel Aviv: Ministry of Defense, 1966), 236, 239; Shtigman, *Me'atsmaut lekadesh,* 199–201.
3. The IAF received its first US-supplied combat aircraft, A-4 Skyhawks, in 1968.
4. Ten C-47 Skytrain/Dakotas and three N-2501 IS Noratlas. On the 890th battalion of the Thirty-Fifth Paratrooper Brigade see Ehud Yonay, *Air Supremacy* (Tel Aviv:

Keter Publishing, 1999), 126–127; Shtigman, *Me'atsmaut lekadesh,* 195–199; and Shachan, *Kanfei hanitsahon,* 236.

5. Yonay, *Air Supremacy,* 184. The total figure included 65 Mirage IIICJs; 35 Super Mystère B.2s; 21 Vautour IIA / B / Ns; 33 Mystère IVAs, and 51 first-generation Ouragans. Also see Ze'ev Lakhish and Meir Amitai, ed., *Asor lo shaket: Prakim betoldot heyl ha'avir bashanim 1956–1967* (Tel Aviv: Ministry of Defense, 1995), 436.
6. Merav Halprin and Aharon Lapidot, *Halifat lahats* (Tel Aviv: MoD / Keter, 2000), 43.
7. Egypt had 299 fighter-bombers (102 MiG-21s, 28 MiG-19s, 96 MiG-17s and MiG-15s, and 16 Sukhoi SU-7s), 57 bombers (30 Tu-16s and 27 IL-28s), 58 transports (IL-14s and AN-12s), and 37 helicopters (Mi-6 / 4s). Syria had 61 MiG-21s, 35 MiG-17s and MiG-15s, 2 IL-28 bombers, 5 IL-14 transports, and 10 Mi-6 / Mi-4 helicopters. Jordan had 24 Hawker Hunter fighters and 7 C-47s while Iraq had 32 MiG-21s, 30 MiG-17s and MiG-15s, and 48 Hawker Hunters as well as 11 IL-28 and 10 Tu-16 bombers. Lakhish and Amitai, *Asor lo shaket,* 438–439.
8. In Sinai: Al-Arish, Bir Gifgafa, Bir Tmade, Jabal Libni; on the Suez Canal: Fayid, Kibrit, Abu-Swer; in the Delta: Inshas, Cairo-West, Cairo-international, Beni Suef, Helwan, Mansoura; in distant areas: Al-Minya, Luxor, Guardaqa, Bilbays, and Ras Banas. Danny Shalom, *Kera'am beyom bahir* (Baavir: Aviation Publications, 2002), 221–462.
9. The Syrian air bases were at Al-Dumayr, Damascus, Marj Ruhayyil, Sayqal, and Tiyas (also known as T4). Shalom, *Kera'am beyom bahir,* 473–518.
10. Yonay, *Air Supremacy,* 184.
11. IAF Headquarters, *50 lemilhemet sheshet hayamim* (Tel Aviv, 2017), 235.
12. Lt. Col. Moti Havakuk, Israeli Air Force chief historian, email communication, May 12, 2018; Lakhish and Amitai, *Asor lo shaket,* 68.
13. Lakhish and Amitai, *Asor lo shaket,* 68.
14. It was persistently misidentified even by reputable historians as *Durendal;* see, Michael B. Oren, *Six Days of War: June 1967 and the Making of the Modern Middle East* (New York: Oxford University Press, 2002), 174.
15. BG (Ret.) Yeshayahu ("Shaike") Bareket, interview, Tel Aviv, March 15, 2016.
16. Some of the major aircraft destroyed were 90 MiG-21s, 20 MiG-19s, 75 MiG-17s, 30 Tu-16 medium bombers, 27 IL-28 light bombers, and 12 Sukhoi-7 fighter-bombers.
17. Author Edward Luttwak was a beneficiary when in the Upper Galilee front line he came under artillery fire but saw no hostile aircraft.
18. Sixty-one airframes were produced locally, as were the SNEMCA Atar 9C engines, by cutting metal to print; on October 7, 1969, the Swiss expelled Israeli military attaché Colonel Zvi Alon after Alfred Frauenknecht was arrested on September 23, 1969, on the accusation of having stolen Atar 9C blueprints from the Sulzer plant in Winterthur.
19. An additional air force technical school was established in Be'er Sheva in 1976.
20. IAF Headquarters, *50 lemilhemet sheshet hayamim,* 235.
21. IAF Headquarters, *50 lemilhemet sheshet hayamim,* 278.

22. Defense Technical Information Center, Technical Report AFFDL-TR-77-115 (December 1977), http://www.dtic.mil/dtic/tr/fulltext/u2/c016682.pdf.
23. The best account in English is Oren, *Six Days of War,* 171. The first Soviet antiaircraft missile, the S-25 Berkut, was never exported.
24. Chris Hobson, *Vietnam Air Losses, United States Air Force, Navy and Marine Corps Fixed-Wing Aircraft Losses in Southeast Asia 1961–1973* (Hinckley, UK: Midland Publishing, 2001), 270, 271.
25. Four aircraft were lost to operational accidents. IAF Headquarters, *50 lemilhemet sheshet hayamim,* 281–286.
26. Edward N. Luttwak, *Strategy: The Logic of War and Peace* (Cambridge, MA: Harvard University Press, 2003), 238.
27. Intelligence on the Soviets included notably the secret de-Stalinization speech. See Matitiahu Mayzel, "Israeli Intelligence and the Leakage of Khrushchev's 'Secret Speech,'" *Journal of Israeli History* 32, no. 2 (September 2013): 257–283. On Operation Diamond (Mivtza Yahalom) see Ian Black and Benny Morris, *Israel's Secret Wars: A History of Israel's Intelligence Services* (New York: Grove Press, 2007), 206–209.
28. Jeffrey T. Richelson, ed., "Area 51 Secret Aircraft and Soviet MIGs," National Security Archive, https://nsarchive.gwu.edu/briefing-book/intelligence/2013-10-29/area-51-file-secret-aircraft-soviet-migs.
29. That was also Edward Luttwak's experience as a consultant on Colonel J. Warden's Instant Thunder attack plan for Desert Shield, the 1990 buildup preceding the Desert Storm attack on Iraq, and then for the US Air Force Chief of Staff, General Merrill A. McPeak, during Desert Storm: the intelligence community only offered generalities, whereas the planners needed exact aiming points. Assorted airmen did the work by improvising, for example, interviewing foreign contractors who built Iraq's aircraft shelters.
30. Yeshayahu ("Shaike") Bareket was an IAF fighter pilot, flight school instructor, squadron leader, and chief of air force intelligence, and he would go on to serve as the deputy IDF attaché in Washington, DC, in August 1973. See Yonay, *Air Supremacy,* 188; and Liat Bloombergerand Tali Ben-Yosef, "2 tayasot, 50 shana," *IAF Magazine,* no. 166 (2005).
31. Yeshayahu Bareket, interview, May 20, 2016, Tel Aviv.
32. There was no redundancy in number of pilots versus planes; therefore almost all pilots had to fly a second sortie, some flying a third. Moti Havakuk, Israeli Air Force chief historian, email correspondence, March 27, 2018.
33. Aviem Sella, interview, Herzliya, Israel, August 6, 2016. His brilliant air force career was interrupted by his accidental involvement in the Pollard espionage case.
34. David Ivry, "Keytsad hishmadnu et ma'arakh hataka bemilhemet shlom hagalil" (Fisher Institute for the Study of Strategy, Air, and Space, Publication no. 36, n.d.), 9.
35. 2K12 Kub; NATO reporting name: SA-6 Gainful.
36. Martin Van Creveld, *The Age of Airpower* (New York: Public Affairs, 2011), 230.

37. On a 1991 visit to Czechoslovakia Aviem Sella (a 1982 Heyl Avir commander) met a Czech general who had served in Moscow in 1982. He related that the air war in Lebanon taught his Soviet counterparts that Western technology was superior. Rebecca Grant, "The Bekaa Valley War," *Air Force Magazine* 85 (June 2002): 58–62; Lior Schlein and Noam Ophir, "Shisha yamim," *IAF Magazine* 145 (June 2002).
38. Ivry, "Keytsad hishmadnu," 230.
39. Meir Finkel, "Pituah hama'aneh letiley hakarka-avir uletkifat sdot te'ufa mimilhemet hahatasha lemilhemet yom hakippurim," *Yesodot* 3 (2021): 30.
40. David Ivry, "Hashmadat ma'arakh hataka bemilhemet shlom hagalil," *Maarchot* 413 (2007): 71; Van Creveld, *Age of Airpower,* 229.
41. Danny Shalom, *Ruah refa'im me'al kahir: Heyl ha'avir bemilhemet hahatasha 1967–1970* (Rishon-LeZion, Israel: Ba'avir Aviation and Space Publishing), 1: 98–101.
42. Shalom, *Ruah refa'im me'al kahir,* 1: 411–412, 415–420.
43. Arie Avneri, *Ha'mahalumah* (Tel Aviv: Revivim and Yediot Aharonot, 1983), 18–27.
44. Avinoam Miseznivkov, "Hapalat Piper 033," Sky-High.co.il, January 11, 2021, https://sky-high.co.il/2021/01/11/%d7%94%d7%a4%d7%9c%d7%aa-%d7%a4%d7%99%d7%99%d7%a4%d7%a8-033/; Ze'ev Schiff, *Knafayim me'al Suez* (Haifa: Shikmona Publishing, 1970), 184.
45. Danny Shalom, *Ruah refa'im me'al kahir* (Rishon-LeZion, Israel: Ba'avir Aviation and Space Publishing, 2007), 2: 1126.
46. Schiff, *Knafayim me'al Suez,* 49–51.
47. Shalom, *Ruah refa'im me'al kahir,* 1: 567, 570.
48. "Rooster 53": at 21:00 on December 26, A-4 Skyhawks and F-4 Phantoms attacked Egyptian ground forces to mask the noise of three SA 321 Super Frelon helicopters carrying a force of Nahal paratroopers and Sayeret Matkal commandos who landed very close to the radar installation. By 02:00 on December 27, the radar components had been taken apart for their return by two CH-53s.
49. Dima Adamsky, *Mivtsa Kavkaz* (Tel Aviv: Ma'arachot, 2006); Isabell Ginor and Gideon Remez, *The Soviet-Israeli War 1967–1973: The USSR's Military Intervention in the Egyptian-Israeli Conflict* (New York: Oxford University Press, 2017).
50. To prevent a recurrence no more buildings were targeted, only SAM sites around Cairo.
51. Danny Shalom, *Ruah refa'im me'al kahir,* 2: 853–854.
52. Shalom, *Ruah refa'im me'al kahir,* 1: 546.
53. Shalom, *Ruah refa'im me'al kahir,* 1: 551.
54. Shalom, *Ruah refa'im me'al kahir,* 2; 855.
55. Shalom, *Ruah refa'im me'al kahir,* 2: 859–860.
56. Shalom, *Ruah refa'im me'al kahir,* 1; 224–227.
57. Shalom, *Ruah refa'im me'al kahir,* 2: 944, 953, 980–984.
58. Shalom, *Ruah refa'im me'al kahir,* 2: 970.
59. Shalom, *Ruah refa'im me'al kahir,* 2: 1002.
60. Shalom, *Ruah refa'im me'al kahir,* 2: 999; Yoav Gelber, *Hahatasha: Hamilhama shenish'k'ha* (Modiin: Zmora Bitan, Dvir -Publishing House, 2017), 461.

61. Shalom, *Ruah refa'im me'al kahir,* 2: 1111.
62. Shmuel Gordon, *Shloshim sha'ot beoktober* (Tel Aviv: Ma'ariv Books, 2008), 154–155.
63. Gordon, *Shloshim sha'ot beoktober,* 267.
64. Shimon Golan, *Milhama beyom hakippurim: Kabalat hahahlatot bapikud ha'elyon bemilhemet yom hakippurim* (Moshav Ben-Shemen, Israel: Modan Publishing House and Ministry of Defense, 2013), 374, 375, 378–379.
65. Golan, *Milhama beyom hakippurim,* 300.
66. Gordon, *Shloshim sha'ot beoktober,* 316.
67. Gordon, *Shloshim sha'ot beoktober,* 345.
68. Gordon, *Shloshim sha'ot beoktober,* 342.
69. It has been the United States' experience in recent wars that its superior airpower was outmaneuvered by enemies that adopted low-contrast modes of action—in Afghanistan or Iraq enemies did not reveal themselves until they attacked, or not even then when using explosive devices detonated remotely. In Israel's case that stage did not occur until after its 1982 Lebanon war.
70. Elchanan Oren, *Toldot milhemet yom hakippurim,* Vol. 2 (Tel Aviv: IDF History Department, 2004), Maps Section, Map 41.
71. Oren, *Toldot,* 2: 529–530.
72. Oren, *Toldot,* 1: 531.
73. Oren, *Toldot,* 2: 6.

11. Airpower Restored with a Technological Leap

1. Arie Avneri, Ha'mahalumah (Tel Aviv: Revivim and Yediot Aharonot, 1983), 18.
2. Eitan Shamir interview with BG (Ret.) Israeli Air Force, Aviem Sella, Herzliya, Israel, August 6, 2016.
3. Shmuel Gordon, *Shloshim sha'ot be'oktober* (Tel Aviv: Ma'ariv Books, 2008), 426.
4. Gil Shani, "Yored mahashamayim," *IAF Magazine Online,* October 25, 2004, http://www.iaf.org.il/1424-22879-he/IAF.aspx.
5. Gordon, *Shloshim sha'ot be'oktober,* 428–431.
6. The MQM-105 Aquila UAV had already demonstrated useful capabilities when it was suspended in September 1985 because it failed 21 of 149 performance specifications, many necessarily trivial: not an uncommon way of stopping macroinnovation that (by definition) lacks extant users to defend it.
7. A stripped-down Firebee (Shadmit in the IAF) was used as a target drone by air defense units.
8. "Hatelem hegiu letayeset hactbamio," IAF, http://www.iaf.org.il/3626-4953-he/IAF.aspx.
9. Too modestly for some; Technion graduate Abraham Karem emigrated to the United States to become the "drone father" with the winning Predator design via first his Albatross, then Amber, experiencing bankruptcy before his company's rebirth as a General Atomics acquisition.

10. Dobster continued at the IAI, developing UAVs for the United States (Pioneer and Hunter), the widely exported armed decoy Harpy, and other UAVs. Eyal Birnberg, "Kesher ayin," IAF, http://iaf.co.il/Shared/Library/Controller.aspx?lang=HE&docID=18389&docfolderID=1102&lobbyID=50.
11. Shani, "Yored mahashamayim."
12. Shani, "Yored mahashamayim."
13. "Tayeset hamalatim harishona," IAF, http://www.iaf.org.il/4968-33518-he/IAF.aspx http://www.iaf.org.il/4968-33518-he/IAF.aspx.
14. 9K33 Osa or Romb; NATO reporting name SA-8 Gecko.
15. David Eshel, "New Tactics Yield Solid Victory in Gaza," *Aviation Week & Space Technology,* May 11, 2009.
16. David A. Fulghum and Robert Wall, "Israel Starts Reexamining Military Missions and Technology," *Aviation Week & Space Technology,* August 20, 2006, https://web.archive.org/web/20061218215607/http://www.aviationnow.com/avnow/news/channel_awst_story.jsp?id=news%2Faw082106p2.xml; "Israel sets combat drones against missile launchers in Gaza," *World Tribune,* May 8, 2007, http://www.worldtribune.com/worldtribune/07/front2454229.238888889.html.
17. Amnon Barzilay, "Ta'asiyat avirit pit'ha matos lelo tayas lehashmadat tilim balisti'im," *Globes,* August 6, 2006.
18. Aviel Magnezi and Yoav Zaitun, "Al kanaf hamalat shehitparka ba'avir hutkan rekhiv hadash," *Ynet,* January 29, 2012, http://www.ynet.co.il/articles/0,7340,L-4182254,00.html.
19. "Katbam hadash leheyl ha'avir: ha'kokhav' hamivtsa'I," IAF, November 10, 2015, http://www.iaf.org.il/4427-45608-he/IAF.aspx.
20. On gliding decoys see Meir Finkel, "Binyan hako'ah lemivtsa 'artsav 19' (1973–1982)," *IDF Journal Bein Ha'Ktavim,* no. 20–21 (2021): 105; and Martin Van Creveld, *The Age of Airpower* (New York: Public Affairs, 2011), 230.
21. Gordon, *Shloshim sha'ot be'oktober,* 427.
22. Finkel, "Binyan hako'ah," 106–109.
23. David Ivry, "Hashmadat ma'arakh hataka bemilhemet shlom hagalil," *Maarchot* 413 (2007): 71.
24. Gordon, *Shloshim sha'ot be'oktober,* 428.
25. Finkel, "Binyan hako'ah," 106–109.
26. Finkel, "Binyan hako'ah," 110–111.
27. Gordon, *Shloshim sha'ot be'oktober,* 282.
28. Eitan Shamir interview with BG (Ret.) Israeli Air Force, Aviem Sella, Herzliya, Israel, August 6, 2016.
29. Gordon, *Shloshim sha'ot be'oktober,* 91–92.
30. Finkel, "Binyan hako'ah," 94.
31. Ivry, "Hashmadat ma'arakh hataka," 70.
32. Finkel, "Binyan hako'ah," 94.
33. Ivry, "Destroying the Syrian SAM array," 69.

34. Eitan Shamir interview with BG (Ret.) Israeli Air Force, Aviem Sella, Herzliya, Israel, August 6, 2016.
35. Gordon, *Shloshim sha'ot be'oktober,* 458; Ivry, "Hashmadat ma'arakh hataka," 70.
36. Sella, interview with the authors.
37. Ivry, "Hashmadat ma'arakh hataka," 71; Van Creveld, *Age of Airpower,* 230.
38. Uri Milstein, "Efekt ha'artsav: Kakh hishmida yisrael et tiliey hasurim ve'et hadoktrina hasoviyetit," *Ma'ariv,* June 4, 2016.
39. Michael Bar Zohar and Nissim Mishal, *No Mission is Impossible* (New York: HarperCollins, 2015), 201.
40. Finkel, "Binyan hako'ah," 109.
41. Avneri, *Ha'mahalumah,* 60–61.
42. Finkel, "Binyan hako'ah," 98.
43. Finkel, "Binyan hako'ah," 101.
44. A year earlier, on June 7, 1981, Sella was chief of operations when the IAF destroyed Iraq's Osirak nuclear reactor in Operation Opera.
45. Finkel, "Binyan hako'ah," 112–113.
46. Kraus and his three team members, Amnon Yoge, Izhak Ben Israel, and Zvi Lapidot, were awarded the prestigious Israel's Defense Prize.

12. Elite Units

1. Boaz Zalmanovitz, "Hakamat kohot meyuhadim belohama nemukhat atsimut," *Ma'arachot,* no. 369 (February 2000): 32–35.
2. For a practitioner's study of Israel's military geography see Yigal Allon, *Masakh shel hol* (Tel Aviv: Hakibutz Hameuchad, 1959), 52–82.
3. Shimon Peres, *Hashalav haba* (Tel Aviv: Am Hasefer, 1965), 9–15.
4. Yehuda Wallach, ed., *Atlas Carta letoldot medinat yisrael—shanim rishonot 1940–1948* (Jerusalem: Carta, 1978), 113.
5. Ze'ev Drory, *Israel's Reprisal Policy, 1953–1956: The Dynamic of Military Retaliation* (London: Frank Cass, 2005), 65.
6. Drory, *Israel's Reprisal Policy.*
7. Shimon Golan, *Hot Border–Cold War* (Tel Aviv: Ma'arachot Publisher, 2008), 308.
8. Drory, *Israel's Reprisal Policy,* 96–101.
9. Moshe Dayan, *Avney derekh* (Tel Aviv: Idanim & Dvir 1976), 159.
10. The IDF had a "minorities unit" of Druze and Circassian volunteers. But in 1956, the Druze leaders opted for male conscription on an equal footing with the Jews.
11. A second attempt to attack the village on January 28–29 also failed; see Drory, *Israel's Reprisal Policy,* 101.
12. Drory, *Israel's Reprisal Policy,* 100.
13. Michael Bar Zohar and Eitan Haver, *Sefer hatsanhanim* (Tel Aviv: A' Levin-Epstein Publishers, 1969), 60.

14. Cited in Shabtai Teveth, *Moshe Dayan: Biografia (Moshe Dayan: A Biography)* (Tel Aviv: Shocken, 1971), 384.
15. In 1951, 62.1 percent of the conscripts were post-1948 immigrants. Drory, *Israel's Reprisal Policy,* 85.
16. Bar Zohar and Haver, *Sefer hatsanhanim,* 63.
17. Dayan had tried to establish a special unit in 1952 when head of Southern Command, Sayeret 30, which did not do well and was soon disbanded. Uri Milstein, *Milhmot hatsanhanim* (Tel Aviv: Ramdor, 1968), 13.
18. Ariel Sharon was subsequently paratrooper brigade commander and head of Southern Command before retiring in July 1973. Recalled to duty that October, he commanded the 143rd Armored Division that crossed the Suez Canal. Minister of defense during the 1982 Lebanon War, he later served as prime minister.
19. Teveth, *Moshe Dayan,* 366.
20. On Unit 101: Mission 101 was Charles Orde Wingate's guerrilla unit during his Ethiopian campaign. Simon Anglim, *Orde Wingate and the British Army: 1922–1944* (London: Routledge, 2015), 124.
21. Benny Morris, *Milhmot hagvul shel yisrael 1949–1956* (Tel Aviv: Am Oved / Afikim Library, 1996), 411–413.
22. Mivtza Shoshana was named after Shoshana Kanias, murdered with her brother and mother two days earlier on October 12, 1953. Efraim Lapid, "Ha'shoshana' shema'adifim lishko'ah," *IsraelDefense,* October 14, 2014, https://www.israeldefense.co.il/content/%D7%94%D7%A9%D7%95%D7%A9%D7%A0%D7%94-%D7%A9%D7%9E%D7%A2%D7%93%D7%99%D7%A4%D7%99%D7%9D-%D7%9C%D7%A9%D7%9B%D7%95%D7%97-%E2%80%93-%D7%A4%D7%A2%D7%95%D7%9C%D7%AA-%D7%A7%D7%99%D7%91%D7%99%D7%94
23. Dayan, *Avney derekh,* 115.
24. Morris, *Israel's Border Wars,* 291 and 448.
25. The future general Mordechai Gur published an account of his days as a young captain in *Company D: The Story of a Paratroopers Company* (Tel Aviv: Ministry of Defense, 1977).
26. April 7, 1954, against Husan; May 9 at Khirbet Ilin; May 27 at Khirbet Jimba, June 28 at Azzoun; August 1 near Jenin; August 13 at Shiekh Madhkur (all in Jordan); April 3, 1954, near Gaza; and August 15 at Bi res Saka (in the Gaza Strip).
27. See Milstein, *Milhmot hatsanhanim;* Arie Avnery, *Pshitot hatagmul* (Tel Aviv: Sifriat Hamachon, 1966); and Bar Zohar and Haver, *Sefer hatsanhanim.*
28. Operation Eye for an Eye, on July 10, 1954, against an Egyptian fort in Gaza.
29. Bar-Zohar and Haver, *Sefer hatsanhanim,* 89–90.
30. In June 1982 MGs A. Ben-Gal and Uri Simchoni, BG Yossi Ben Hanan, and Major Meir Dagan (future head of Mossad), together with Edward Luttwak and two sergeants, drove north to Byblos (Jbeil) fifty kilometers beyond Israeli lines in Lebanon, and then another thirty kilometers east just to observe, coming within pistol range of Syrian troops.

31. Yotam Amitai and Tamar Barash, "Yehidot meyuhadot betsahal ba'avar uvahoveh: Nitu'ah metahim tsavi'im-hevrati'im," *Ma'arachot* no. 411 (February 2007): 15–22.
32. Ilan Kfir and Ben Kaspit, *Ehud Barak: Hayal mispar 1* (Tel Aviv: Alpaha Tikshoret, 1998), 39–47, 147–152, 246.
33. "Hativat hakomando shel tsahal yotset laderekh," *Mako,* December 27, 2015, https://www.mako.co.il/news-military/security-q4_2015/Article-42880b14824e151004.htm
34. Keren Hellerman, "Ma meyuhad bayehidot hameyuhadot," *Between the Arenas* 3 (2007): 21–29.
35. Idan Soncino, "Esh mitahat la'adama: Kakh yehidot haHIR mitamnot belohama tat-karka'it," *Mako,* July 7, 2012, https://www.mako.co.il/pzm-magazine/Article-a7f54f1dbb49831006.htm.
36. Amos Harel and Gili Cohen, "Bli tokhniyot, imunim vetsiyd: Kakh hitmoded tsahal im haminharot," *Ha'aretz,* October 17, 2014.
37. Yoav Limor, "Anshey harefa'im," *Israel Today,* August 20, 2020.
38. Arnon Schwartzman, "Shinuy be'itur halohamim leyehidat tsahal hehadasha: 'He kan kedey lehisha'er,'" *Mako,* May 5, 2021, https://www.mako.co.il/pzm-soldiers/Article-64cc39064bb2971027.htm.
39. Tal Ram Lev, "Rahfan lekhol mem-mem vehafalat esh mehira yoter: Kakh year'eh he'atid shel kohot hayabasha," *Ma'ariv Online,* March 18, 2021, https://www.maariv.co.il/news/military/Article-828550.

13. Military Entrepreneurs and Special Forces

1. Judy Baumel, "Tzava'ato shel Uri Ilan," *Iyunim Be'Tkumot Israel* 15 (2005): 209–238, https://in.bgu.ac.il/bgi/iyunim/15/judy.pdf.
2. Lior Brichta and Eyal Ben-Ari, "Organizational Entrepreneurship and Special Forces: The First Israeli Helicopter Squadron and the General Staff Reconnaissance Unit (Sayeret Matkal)," in *Special Operations Forces in the 21st Century: Perspectives from the Social Sciences,* Jessica ed. Glicken Turnley, Kobi Michael, and Eyal Ben-Ari (Abingdon, UK: Routledge, Cass Military Studies, 2017), 213–214.
3. Avner Shor, *Hotseh gvulot: Sayeret matkal umeyasda Avraham Arnan* (Modi'in, Israel: Kineret Zamor Bitan Dvir Publishing, 2008), 76–77.
4. Shor, *Hotseh gvulot,* 90–92. See also Amnon Jackont, *Meir Amit: Ha'ish vehamosad* (Tel Aviv: Yediot Books Publishing, 2012), 94.
5. Lior Brichta, Razi Efron, and Pinhas Yehezkeally, *Yehidat 101: Ee al saf haka'os* (Be'er Sheva: DNA T.E.C.I., 2012).
6. Shor, *Hotseh gvulot,* 97–101.
7. Brichta, Efron, and Yehezkeally, *Yehidat 101,* 22–23; Shor, *Hotseh gvulot,* 102–103.
8. Shor, *Hotseh gvulot,* 106–107.
9. Shor, *Hotseh gvulot,* 121–123.

10. Shor, *Hotseh gvulot,* 108–118. See also Moshe Zonder, *Sayaret Matkal* (Jerusalem: Keter Publishing House), 24.
11. Shor, *Hotseh gvulot,* 118–119.
12. Haim Laskov, who had served in the Jewish Brigade in World War II, rose to chief of staff in 1958 after Dayan.
13. Shor, *Hotseh gvulot,* 112. See also Uri Yarom, *Kanaf renanim* (Tel Aviv: Ministry of Defense, 2001), 224.
14. Hanoch Bartov, *Dado—48 shana ve'od 20 yom* (Tel Aviv: Sifriyat Ma'ariv, 1978), 77.
15. For "requisitioning," see Uri Ben-Ari, *"Nua nua! Sof." Hama'avak al derekh hashiryon* (Tel Aviv: Ma'arachot, 1998); Shor, *Hotseh gvulot,* 114–116; and Zonder, *Sayaret Matkal,* 24–25.
16. Yigal Sheffy, *Hatra'a bemivhan* (Tel Aviv: Ma'arachot, 2008), 71.
17. Shor, *Hotseh gvulot,* 131.
18. Shor, *Hotseh gvulot,* 135.
19. Ofer Drori, "Mivtsa'ey haluts vashrakrak," http://www.gvura.org/343185-; Yosef Castel, *Hayay—Yoske Castel (My Life—Yoske Castel)* (Tek Aviv: Yoske Castel, 2010), 88–89.
20. Amos Gilboa, *Mar modi'in: Areleh, aluf Aharon Yariv, rosh aman* (Tel Aviv: Miskal Yediot Books, 2013), 111–115. See also Shor, *Hotseh gvulot,* 181–186.
21. Shor, *Hotseh gvulot,* 203. See also Gilboa, *Mar modi'in,* 111.
22. Brichta and Ben-Ari, "Organizational Entrepreneurship."
23. Yizhak Shteigmann, "The Introduction of Helicopters into the Israeli Air Force 1948–1958," *Cathedra for the History of Eretz Israel and Its Yishuv* 53 (1989): 131–148.
24. Meir Amitai, *Ad 124: Tayeset hamesokim* (Tel Aviv: Zamora Bitan Publishing, 1990), 24–25.
25. Yarom, *Kanaf renanim,* 153.
26. Brichta and Ben-Ari, "Organizational Entrepreneurship," 219.
27. Shor, *Hotseh gvulot,* 212.
28. Shor, *Hotseh gvulot,* 215. In 1967 Arnan established a nonprofit, Amutat Misdar Dorshei Hatov, the "Order of Those Who Want To Do Good," which recruits unit veterans for social work.
29. Amir Oren, "Mi natan et hapkuda," *Ha'aretz,* April 8, 2012. Also see Avner Shor, *Tsevet Itamar* [Team Itamar]*: Sayeret Matkal, the People, the Operations, the Atmosphere* (Jerusalem: Keter Publishing, 2003), 65–72; and Zonder, *Sayaret Matkal,* 67.
30. Betzer served in the IDF from 1964 to 1986, starting in the Paratroopers Brigade before moving to the *Sayeret Matkal*. Muki Betzer, *Lohem hashay* (Jerusalem: Keter Books Ltd., 2015), 345–365.
31. Adam was the highest-ranking IDF officer ever killed in combat, on June 10, 1982, near Damour south of Beirut, when sheltering from artillery fire in an uncleared building. Betzer, *Lohem hashay,* 345–365.
32. After retirement, Yiftach Spector inspired research on automatic target recognition, published the prize-winning novel *Ram uvarur* (Rishon Le'Tzion, Israel: Miskal Yediot Ahronot, 2008), and became a peace activist.

33. Spector, *Ram uvarur,* 288ff. Major General Rafael ("Raful") Eitan was deputy chief of general staff, becoming chief in 1978, serving until 1983. Spector, *Ram uvarur,* 290–291.
34. Spector, *Ram uvarur,* 287–289.
35. Spector, *Ram uvarur,* 358.
36. Avichai Becker, "Ptsatsat testosterone," *Ha'aretz,* August 13, 1999.
37. Revealed by BG (Ret.) Gal Hirsh, the unit commander. Hirsh and Shaldag were decorated. Gal Hirsh, *Sipur milhama sipur ahava* (Tel Aviv: Miskal Yedioth Ahronoth and Chemed Books, 2009), 113–123.
38. Shai Levy, "Hakomando ha'aviri: Hamivtsa'im hagdolim shel yehidat shaldag," *Mako,* November 7, 2012, https://www.mako.co.il/pzm-magazine/army-stories/Article-2e8e84ed23ada31006.htm.
39. See Ofer Shelach and Yoav Limor, *Shvuyim bil'vanon* (Tel Aviv: Miskal Yedioth Ahronoth and Chemed Books, 2009), 249–250.
40. Yossi Melman and Dan Raviv, "Hashmadat hakur hasuri: Hasipur shelo supar," *Ha'aretz,* August 3, 2012.

14. The Armored Corps

1. David Eshel, *Chariots of the Desert: The Story of the Israeli Armoured Corps* (London: Brassey, 1989), 5.
2. The British soldiers, Mike Flanagan and Harry McDonald, drove the tanks some 100 kilometers from Haifa to a Haganah hideout. The two remained in Israel; Flanagan's grandson became a major in the armored corps. Avi Eliyahu, "Sipuro hamadhim shel hatank harishon betsahal," *Mako,* June 2, 2014, at: https://www.mako.co.il/pzm-units/armored-corps/Article-0c5a3f9195c5641006.htm; Shabtai Teveth, *The Tanks of Tammuz* (London: Weidenfeld & Nicolson, 1969). There may have been a second Sherman. Yehuda Wallach, "Hitpathut hamah'shava hashiryonit betsahal," *Ma'arachot,* no. 197 (1961): 15.
3. Teveth, *Tanks of Tammuz,* 41.
4. Eshel, *Chariots of the Desert,* 25.
5. Eventually the IDF operated four grades of M48 Pattons: Magach 1, the original model with petrol engine and 90 mm gun; Magach 2: M48A2/E48A; Magach 3: modified in Israel to carry a 105 mm L7A1 gun, an Urdan low-profile commander's cupola, a new communication suite, and a (AVDS-1790-2A 750 hp) diesel engine; and the final upgrade, *Magach 5,* with upgraded AVDS-1790-2D and transmission.
6. Minute by S. L. Egerton, May 14, 1970, TNA, FCO 17/1303.
7. Designed by the Soviets based on German technology captured in 1945 but almost ignored in the West, as with the helmet-mounted displays that equipped MiG-29s and were overlooked by the USAF even when East German MiG-29s fell into Western hands. But the IDF experts took them very seriously, developing the design into what is now the F-35's major advance.

8. Edward Luttwak saw a battalion of what he believes were original Shermans on their way to the Golan Heights on June 9, 1967.
9. L-33 Ro'em with Soltam 155 mm self-propelled howitzers for mobile artillery fire; M-50s with the French 155 mm howitzer mounted at the back of the M4A4 Sherman hull; Makmat 160 mm mortars for even heavier if shorter-range fire; MAR-240 mobile launchers for thirty-six heavy 240 mm bombardment rockets; MAR-290 mobile launchers for four 290 mm ground-to-ground rockets; Ambutank, an ambulance on the original Sherman chassis; Eyal Sherman with an observation pod elevatable to twenty-seven meters; Kilshon, using the hull of the Sherman M-51 with AGM-45 Shrike antiradiation missile launchers; the Sherman Morag mine flail vehicle; and the Trail Blazer (Gordon) Recovery / engineering vehicle on HVSS-equipped M4A1s.
10. When Tal was the Armored Corps Commander, he once served as a tank gunner (under the command of one of his subordinates) in a firefight with Syrian tanks; he destroyed a Syrian tank, but his own tank was also damaged.
11. Teveth, *Tanks of Tammuz,* 60.
12. At thirty-four to thirty-six tons combat weight, the Sherman chassis could not absorb the recoil of the high-velocity L.7 A1. 105 mm gun fitted to the fifty-two-ton Centurion. Upgraded Israeli Sherman M50s were fitted with the AMX-13 75 mm gun and Sherman M51s with a French medium-velocity (800m / s) 105 mm gun firing hollow-charge rounds, but some units still had Shermans with older 76mm high-velocity guns or even the original 75 mm guns.
13. Teveth, *Tanks of Tammuz,* 55–57, 63–67, 71–73.
14. Born in 1933, Held earned a PhD in physical chemistry (ultraviolet spectroscopy) at TU-Muenchen in 1959. He worked at Messerschmidt-Boelkow-Blohm, the postwar German tank house, on initiation, detonation, fragmentation, cylinder expansion, penetration, and metal acceleration. Florian Bouvenot, "The Legacy of Manfred Held with Critique" (Monterey, CA: Naval Post-Graduate School, 2011), xxiii, 2–5; Norbert Eisenreich, "Manfred Held, A Life Devoted to Explosive Science," *Propellants, Explosives, Pyrotechnics* 41, no. / 1 (2016): 7.
15. "Dr. Manfred Held Memorial Presentation: A Celebration," *National Defense Industrial Association,* http://www.dtic.mil/ndia/2011ballistics /DrManfredheldMemorial.pdf.
16. When the ERA boxes were exported after the 1982 war, Held earned substantial royalties, as Tal had secured his rights.
17. Tom Cooper and Yaser el-Abed, "Syrian Tank Hunters in Lebanon, 1982," *ACIG, The Middle East Data Base,* September 23, 2003, https://web.archive.org/web /20080321015417/http://www.acig.org/artman/publish/article_279.shtml.9
18. Soeren Suenkler and Marsh Gelbart, *IDF Armored Vehicles: Tracked Armour of the Modern Israeli Defense Forces (IDF)* (Erlangen: Tankograd Publishing, Verlag Jochen Vollert, 2006), 10.
19. David Eshel, "Did Merkava Challenge its Match?," *Armor* 11, no. 1 (January–February 2006): 44–46.

20. Hanan Greenberg, "Why Did the Armored Corps Fail in Lebanon 2006," *Ynet News,* August 30, 2006; Suenkler and Gelbart, "IDF Armored Vehicles."
21. Shaul Nagar, "Prof. Manfred Held shepite'ah shiryon re'aktivi halakh le'olamo," *Yad LaShiryon.com,* February 17, 2011.

15. Why the Merkava Is Different

1. S. C. Smith, "Centurions and Chieftains: Tank Sales and British policy towards Israel in the Aftermath of the Six Day War," *Contemporary British History* 28, no. 2 (2014): 219–239.
2. Smith, "Centurions and Chieftains," 5ff.
3. Saul Bronfeld, "Albion habogdanit: Hatik," *Shiryon,* no. 37 (March 2011): 26–31, https://yadlashiryon.com/wp-content/uploads/2017/02/%D7%92%D7%9C%D7%99%D7%95%D7%9F-%D7%9E%D7%A1%D7%A4%D7%A8-37-1.pdf
4. Israel Tal, interview with Mordechai Bar-On and Pinchas Ginosar, *Iyonim Betkomat Yisrael,* Ben Gurion University, Vol. 10, 2000, 66, http://in.bgu.ac.il/bgi/iyunim/10/2.pdf.
5. The project transformed the Ordnance Corps Masha (*Merkaz Shikum veAkhzaka,* "Repair and Maintenance Center") facilities into the assembly line of the Merkava.
6. Tal, interview with Bar-On and Ginosar, 72.
7. Almost but not always—approximately 200 Israeli tanks damaged on the Sinai front the first four days of the October 1973 war, a handful lost in the battle for Suez the last day of the war, and three Patton M48 tanks damaged in the 1982 Sultan Ya'akub battle remained in enemy hands. The tanks lost in 1982 carried reactive-armor plates and were sent to the USSR, which reverse-engineered copies. All Russian tanks since have carried gradually improved versions of that Israeli technology.
8. Tal, interview with Bar-On and Ginosar, 78.
9. Based on Tal interview. Also see Patrick Wright, *Tank* (New York: Penguin Books, 2003), 323–366.
10. "Kol ma sheratsita lada'at al hatokhnit harav-shntatit Gideon," IDF, July 26, 2015, http://www.idf.il/1133-22449-he/Dover.aspx. Since 1985, the number of tanks has fallen by 75 percent, the number of aircraft by 50 percent, and the number of UAVs has increased by 400 percent. See Amir Rapaport, "The New Multi-Year Plan of the IDF and the Agreement with Iran," *Israel Defense,* September 9, 2015. From 1989 to 2015 the number of active-duty armored brigades was reduced from six to four and the number of reserve brigades was reduced from eighteen to eight. The IDF's entire remaining fleet of M-60s, Centurions, and Merkava Mk 1s and Mk 2s has been decommissioned, leaving only about 1,500 Merkava Mk 3s and Mk 4s in the order of battle.
11. Provided by Dr. Eado Hecht, with whom the battalion commander spoke in October 2006.

12. Nicholas Blanford, *Warriors of God: Inside Hezbollah's Thirty-Year Struggle against Israel* (New York: Random House, 2011), 406–407; "Hipagut usridut tankim bemilhmot yisrael," *Shiryon,* no. 24 (October 2006): 55, https://yadlashiryon.com/wp-content/uploads/2017/02/%D7%92%D7%9C%D7%99%D7%95%D7%9F-%D7%9E%D7%A1%D7%A4%D7%A8-24-1.pdf.
13. Col. Benny Michaelson, "Hashiryon bemilhemet levanon hashniya," *Shiryon,* no. 30 (December 2008); 26–33, esp. 33.
14. Hanan Greenberg, "Why Did the Armored Corps Fail in Lebanon 2006," *Ynet News,* August 30, 2006; Soeren Suenkler and Marsh Gelbart, "IDF Armored Vehicles," *Tracked Armour of the Modern Israeli Defense Forces* (Erlangen: Tankograd Publishing, Verlag Jochen Vollert, 2006).
15. Trophy System (ASPRO: Armored Shield Protection—Active).
16. Nadav Paz, "Hashorashim shel hamigun letankim—me'il ru'ah," *Israel Defense,* https://tinyurl.com/kh2vrmo.
17. Paz, "Hashorashim"; Yiftach Klinman, "Me'il ru'ah—neshek hahakhra'a," *Ma'arachot,* no. 450 (2013): 72–73.
18. "Rafael and Elta Unveil the Trophy Active Protection System," Rafael Advanced Defense Systems Ltd., http://www.Rafael.co.il/Marketing/192-964-en/Marketing.aspx.
19. Paz, "Hashorashim."
20. Amir Buchbut, "Me'il ru'ah shel tsahal hofekh mivtsa'i lehagana al tankim," *NRG,* August 6, 2009.
21. "Trophy System Thwarts Missile Fired at IDF Tank," IDF, https://www.idfblog.com/blog/2011/03/01/windbreaker-thwarts-missile-fired-at-idf-tank/.
22. Ron Ben Yishai and Elior Levi, "Muganim im me'il ru'ah: Zuha til sheshugar al tank," *Ynet,* March 20, 2011, https://www.ynet.co.il/articles/0,7340,L-4044859,00.html.
23. Florit Shoychat, "Hativa shlema im me'il ru'ah: Bemilhama nuchal lehagi'a yoter amok veyoter rahok," IDF, http://www.idf.il/1133-16373-he/Dover.aspx
24. Or Heller, "Hativa 7 kalta et tank hamerkava siman 4," *Israel Defense,* October 28, 2014, https://tinyurl.com/ms7xn5q.
25. Michael B. Kim, "The Uncertain Role of the Tank in Modern War: Lessons from the Israeli Experience in Hybrid Warfare" (Land Warfare Paper No. 109, Institute of Land Warfare, Association of US Army, Arlington, VA, July 2016), 15.
26. Amir Buchbut, "Hitsil 15 tankim: Kipat barzel shel hashiryon be'aza," *Walla News,* June 6, 2014; Yossi Yehoshua, "Shuvam shel hatankim, hahatslaha shel me'il ru'ah," *Ynet,* July 27, 2014, https://www.ynet.co.il/articles/0,7340,L-4550567,00.html.
27. Barbara Opall-Rome, "Israel to Equip Troop Carriers with Trophy APS," *Defense News,* January 28, 2016.
28. "Trophy," Rafael Advanced Defense Systems Ltd., http://www.Rafael.co.il/Marketing/281-963-en/Marketing.aspx.
29. "Trophy", Rafael Advanced Defense Systems Ltd.
30. "Iron Fist Active Protection System (APS)," *Defense Update,* https://defense-update.com/products/i/iron-fist.htm.

31. "IDF Approves Acquisition of IMI's Iron Fist Active Protection Systems for Namer AIFVs," *Defense Update,* http://defense-update.com/features/2009/june/idf_aps_090609.html.
32. "Rafael Proposes its Trophy Active Protection System to U.S. Army," *Army Recognition,* http://www.armyrecognition.com/october_2014_global_defense_security_news_uk/Rafael_proposes_its_trophy_active_protection_system_to_u.s._army.html.
33. Yaakov Lappin, "US Army Selects Israel Military Industries for APC Active Protection System," *Jerusalem Post,* June 7, 2016.
34. Judah Ari Gross, "US Army Inks $193 Million Deal to Buy Israeli Tank Defense System," *Times of Israel,* June 26, 2018, https://www.timesofisrael.com/us-army-inks-193-million-deal-to-buy-israeli-tank-defense-system/.
35. Colton Jones, "British Army Selects Israel's Trophy Active Protection System for Its New Tanks," *Defense Blogs,* June 27, 2021, https://defence-blog.com/british-army-selects-israels-trophy-active-protection-system-for-its-new-tanks/.
36. Meirav Ankori, "Pagaz hakalanit shel hata'asiya hatsva'it yikanes leshimush mivtsa'i betsahal," *Globus,* July 28, 2009, https://www.globes.co.il/news/article.aspx?did=1000484790; "120mm APAM-MP-T, M329 Cartridge," IMI Systems.
37. Daniela Bokor, "Proyekt tsva yabasha digitali memshikh lehoshit yadayim," IDF, https://www.idf.il/1133-11523-he/Dover.aspx.
38. "Hahatsav yahalif et hahalulan," *Israel Defense,* http://bit.ly/2mHIa4N.
39. Amir Bar Shalom, "Hidushey TAAS betsuk Eitan," *New Tech Military Magazine,* http://bit.ly/2l8OYft.
40. Rafi Rubin, "Kalanit: Pagaz 120 mm rav-shimushi vehadshani," *Shiryon,* no. 37 (March 2011), http://www.yadlashiryon.com/show_item.asp?levelId=64566&itemId=2578&itemType=0).
41. "120mm APAM-MP-T, M329 Cartridge," IMI Systems.
42. Yuval Azulai, "Na lehakir: Ma'arkhot hahagana hamitkadmot beyoter betsahal," *Globes,* 7 August 2014.
43. "Hakalanit porahat," *Fresh,* http://www.fresh.co.il/vBulletin/showthread.php?t=475282&highlig; "120mm M329 APAM-MP-T Tank Cartridge," IDI, http://www.imi-israel.com/vault/documents/120mm%20m329%20apam-t_nt-001_draft_0034.pdf.
44. Nir Segal, "Heyl hashiryon hisel lemala me-500 mehablim," *Mako,* August 28, 2014.
45. Yehali Sa'ar, "Kalanit baderekh el hativot hashiryon," *Fresh,* http://www.fresh.co.il/vBulletin/showthread.php?t=475282&highlig post3786292; "Hatzav Will Replace the Halulan," *Israel Defense,* November 23, 2011.
46. Shay Levi, "Lo mashirim sikun: Hahimush hehadash shel heyl hashiryon," *Mako,* September 23, 2014.
47. Gili Cohen, "Tahkir tsahal: NAGMASH hayaley golani haya taku'a ka'asher safag esh," *Ha'aretz,* July 21, 2014, https://www.haaretz.co.il/news/politics/2014-07-21/ty-article/.premium/0000017f-e230-d75c-a7ff-febde6a10000.
48. "Bekarov: Gam HaNAMERim yetsuydu beme'il ru'ah," Ministry of Defense, http://www.mod.gov.il/Defence-and-Security/articles/Pages/31.1.16.aspx.

16. Units 8200 and 81

1. Geoffrey Ingersoll, "The Best Tech School on Earth Is Israeli Army Unit 8200," *Business Insider,* August 13, 2013.
2. Or Hirshaoga, "Profil hahitek hayisraeli: Bahur tsa'ir yotseh yehidat tsahal tekhnologit o lohemet veboger universita," *The Marker,* March 10, 2013.
3. "Yehidat 8200," IDF, http://bit.ly/2oCsfIW.
4. "Hasiha hasodit shel Nasser veHussein," Israel Intelligence Heritage and Commemoration Center, http://malam.cet.ac.il/ShowItem.aspx?ItemID=b0351703-4eed-45ee-b3b4-bba7d56d6ef3&lang=HEB.
5. Nir Dvori, "Biladi: Hakolot meEntebbe neh'safim," *Mako,* June 16, 2016, https://www.mako.co.il/news-channel2/Channel-2-Newscast-q2_2016/Article-bd44a8ae94a5551004.htm.
6. "Israel Foiled Plane Terror Plot in Australia," *BBC News,* February 22, 2018, https://www.bbc.com/news/world-australia-43149722.
7. "Snowden *Der Spiegel* Interview," Der Spiegel, http://web.archive.org/web/20130708030634/http://cryptome.org/2013/07/snowden-spiegel-13-0707-en.htm.
8. Cybersecurity is under the responsibility of other organizations in the IDF.
9. Oded Yaron, "Hokrey bitahon: Stuxnet pa'il me'az 2007," *Ha'aretz,* February 26, 2013.
10. Dudi Cohen, "Iran moda: Harbeh meyda avad, yesh kesher leStuxnet," *Calcalist,* May 29, 2012.
11. "A Complex Malware for Targeted Attacks," Laboratory of Cryptography and System Security, https://www.crysys.hu/skywiper/skywiper.pdf.
12. Cohen, "Iran moda."
13. Oded Yaron, "Iran moda: Havirus ganav me'itanu harbeh meyda veshibesh yetsu haneft," *Ha'aretz,* May 30, 2012.
14. Cohen, "Iran moda."
15. "Complex Malware."
16. Ehud Keinan, "Yotsrey virus lehava gormim lo lehitabed," *Ynet,* June 8, 2012, https://www.ynet.co.il/articles/0,7340,L-4239985,00.html.
17. Cohen, "Iran moda."
18. See Amir Rapaport, "Hashin bet be'idan hasiber: Mabat mebifnim," *Israel Defense,* April 11, 2014.
19. Michael Danieli, "8200: Hakiru et hayehida hamesuveget hakhi gdola betsahal," *Mako,* September 12, 2011, https://www.mako.co.il/pzm-units/intelligence/Article-8547b921354f031006.htm.
20. Danieli, "8200."
21. Danieli, "8200."
22. Danieli, "8200."
23. "Unit 8200," Israel Intelligence Heritage and Commemoration Center, http://malam.cet.ac.il/ShowItem.aspx?ItemID=e44c41b1-5961-40ec-b8bd-9acaa6d2f6d1&lang=HEB.

24. Chief NCO M., interview, Herzliya, June 9, 2016. The position is classified therefore interview could not reveal name.
25. "Unit 8200," Israel Intelligence Heritage and Commemoration Center.
26. "Unit 8200," Israel Intelligence Heritage and Commemoration Center.
27. Israel Parliamentarian Sub-Committee on Military Preparedness and General Scrutiny, "IDF Readiness for War, HUZBIT-63-458," Jerusalem, 2020.
28. "Yehidat 8200," IDF.
29. Yochai Ofer, "Lohamey 8200 hosfim: Kakh hisalnu mehablim betsuk Eitan," *NRG,* October 8, 2014, https://www.makorrishon.co.il/nrg/online/1/ART2/630/560.html.
30. Assaf Gilad and Meir Orbach, "8200 pinat emek silicon: Hayehida hagdola betsahal lomedet la'avod kemo start-up," *Calcalist,* July 1, 2012, https://www.calcalist.co.il /internet/articles/0,7340,L-3575727,00.html.
31. Interview with officers in Unit 8200.
32. Shay Levi, "Ekh po'elet sokhnut harigul hagdola ba'olam?," *Mako,* October 20, 2013, https://www.mako.co.il/pzm-magazine/foreign-forces/Article -f2f62ba54690241006.htm.
33. Ran Dagoni, "Sarvaney 8200 me'amtim te'anot Snowden al shituf pe'ula im haNSA," *Globes,* September 17, 2014, https://www.globes.co.il/news/article.aspx?did =1000972370.
34. Nicky Hager, "Israel's Omniscient Ears," *Le Monde Diplomatique,* September 2010.
35. From interviews with serving officers and NCOs in Unit 8200.
36. Mordechai Naor, "Hamehdal hakaful bemutsav haHermon," *Ha'aretz,* September 16, 2013, https://www.haaretz.co.il/literature/study/2013-09-16/ty-article/.premium /0000017f-f153-da6f-a77f-f95fddec0000.
37. Amnon Lord, "K'tsin hamodi'in shel hareshet hashara: Amru li shemedinat yisrael kvar lo kayemet," *NRG,* November 1, 2013.
38. The term for reserve forces that reinforce the standing army is *Miluim:* "filling in."
39. Interviews with serving Unit 8200 officers and NCOs.
40. Interviews with serving Unit 8200 officers and NCOs.
41. Amos Harel, "Be-8200 meshabhim hasarvan al musariyut akh mevakrim hitnahaguto," *Haaretz,* January 30, 2003, https://www.haaretz.co.il/misc/2003-01-30/ty -article/0000017f-e104-d75c-a7ff-fd8d9e4c0000.
42. Lieutenant Colonel Uri, "Lachtzuv ma'yim mh'asela: shinuy vehishtanut bmanganoney ha'mup byisreal," *Between the Poles: Power Building—Part 2* 7 (2016): 41–59.
43. Interviews with serving 8200 officers.
44. Uri, "Lachtzuv ma'yim mh'asela."
45. From the Ministry of Defense memorial website for fallen soldiers, Izkor: https://www.izkor.gov.il.
46. David Kushner, "The Israeli Army's *'Roim Rachok'* Program Is Bigger Than the Military," *Esquire,* April 2, 2019.
47. In Hebrew *Roim Rachok* can have a dual meaning: seeing far or seeing into the future.
48. Rotem Abrutzky, "Hayalim al hakeshet" (the Hebrew can mean "on the spectrum"), *Israel's Public TV Ka'an 11,* February 16, 2016.

49. Lieutenant General Gadi Eisenkot, former IDF chief of staff, interview, Tel Aviv, March 2021.
50. Abrutzky, "Hayalim al hakeshet."
51. Abrutzky, "Hayalim al hakeshet."
52. See the Roim Rachok program Facebook site: https://www.facebook.com/Roim Rachok/.
53. Kushner, "The Israeli Army's *'Roim Rachok'* Program Is Bigger Than the Military."
54. Anshil Pepper, "Ne'arey Raful: Haproyekt hakhi hevrati shel tsahal hogeg 30," *Ha'aretz,* August 28, 2010, https://www.haaretz.co.il/misc/2010-08-27/ty-article/0000017f-e2f8-d75c-a7ff-fefd2cea0000.
55. Michal Yaakov Yitzchaki, "Miposhe'a lelohem: Hamahapakh shel Lior," *Israel Today,* April 6, 2020, https://www.israelhayom.co.il/article/749351.
56. From 2007 a concentrated effort was made to steer program graduates to enlist in combat units. In 2007, a total of twenty-seven soldiers joined combat units; in 2008 that number doubled, then tripled the next year. In 2009 the first graduate of the program was commissioned as a combat officer. Hanan Greenberg, "Hahazon hitgashem: Ktsin kravi rishon mina'arey Raful," *Ynet,* February 27, 2009, https://www.ynet.co.il/articles/0,7340,L-3677915,00.html. See additional examples from more recent years in Korin Elbaz, "Hana'ar im tikim plili'im hafakh lekatsin mitztayen," *Ynet,* October 13, 2017, https://www.yediot.co.il/articles/0,7340,L-5027777,00.html.
57. See Lilach Lev Ari, Michal Razer, Noa Ben Yosef-Azoulay, and Rinat Adler, "Meniduy vesikun lehakhala veshiluv: Bogrey tokhniyot meyuhadot betsahal mishtalvim ba'ezrahut," *Mifgash: Journal of Social-Educational Work* 24, no. 43 (June 2016): 59–85.
58. See Israel Ministry of Defense: https://www.mod.gov.il/Society_Economy/articles/Pages/25.8.20.aspx
59. "Kakh mifkedet Alon ozeret likto'a et sharsheret hahadbaka," IDF, April 27, 2021.
60. *Kupat Holim* ("Sick Funds"): in addition to this across-the-board public service, Israelis can increase medical coverage by purchasing private health insurance. In 2020, Israel's health system was ranked third most efficient in the world. See "Asia Trounces U.S. in Health-Efficiency Index Amid Pandemic," *Bloomberg.com,* December 18, 2020; and Yoav Zeiton, "Mifkedet haCorona betsahal hehela lifol: Habdika vektiyat hahadbaka tokh 36 sha'ot," *Calcalist,* August 4, 2020, https://www.calcalist.co.il/local/articles/0,7340,L-3843423,00.html.
61. Nir Dvori, "Sayeret Corona: Hapituhim hayisraeli'im shel hatsevet hameyuhad yenats'hu et hanagif?," March 23, 2020, *N12 News,* https://www.mako.co.il/news-military/2020_q1/Article-c3a8c3be3a60171026.htm.
62. Hanan Greenwood, "Hayehida hamesuveget sheshinta et hatmuna," *Israel Today,* April 2, 2021; Ori Berkowitz, "Kakh yehidat modi'in mesuveget guysa lelhilahem bingif haCorona," *Globes,* April 17, 2020.
63. Sophie Shulman, "Yehida ktana, mapats gadol," *Calcalist,* January 7, 2021, https://newmedia.calcalist.co.il/magazine-07-01-21/m01.html.

Conclusion

1. Edward Luttwak, personal recollections (with photographs), Lebanon, 1982.
2. Secretary of Defense Caspar W. Weinberger was unequivocal: "We want . . . Israeli forces completely out of Beirut . . . There are still far too many foreign forces in Lebanon—Syrians, Israelis . . .": "An Interview with Caspar Weinberger," *Washington Post,* September 26, 1982, https://www.washingtonpost.com/archive/opinions/1982/09/26/an-interview-with-caspar-weinberger/48e7fd8d-9063-4e55-920a-0493f8415341/.
3. MG Israel Tal of the Armored Corps would have agreed with the US Marines.
4. Dr. (Lt. Col. Ret.) Eado Hecht, personal communication, Tel Aviv, 2016.
5. The postwar U.K. Royal Commission on Awards to Inventors of 1919 determined that a 1912 proposal received by the War Office for a fighting vehicle armored to resist machine-gun fire, and tracked to crush barbed wire and drive over trenches, was superior to the tank actually produced from 1916. It had been filed, unread. Gray E. Dwyer, "Story of the Tanks," *The West Australian* (Perth WA), August 11, 1924.
6. Gregory C. Allen, "Project Maven Brings AI to the Fight against ISIS," *Bulletin of the Atomic Scientists,* December 21, 2017; Ethan Baron, "Google Backs Off from Pentagon Project after Uproar: Report," *Military.com,* June 3, 2018.

Acknowledgments

EDWARD LUTTWAK: I am grateful to MG Aluf Aharon Yariv z.l., who hired me to expand the geographic scope of IDF intelligence and allowed me to reach the Sinai front in October 1973; MG Uri Simchoni z.l., who invited me to participate in the design of a special operations unit; MG Israel Tal z.l., who allowed me to monitor the development of his Merkava tank; MG Avigdor Ben-Gal z.l., who invited me to join his "superman" foray to northernmost Lebanon; the US Army's Training and Doctrine Command, which invited me to participate in its introduction of maneuver warfare and my own operational-level concept in US military doctrine; and, above all, to the Office of Net Assessment, Office of the Secretary of Defense, US Department of Defense for supporting many geopolitical explorations (including one published as *The Rise of China viz the Logic of Strategy*) and then taking the risk of funding a technology project, Reinventing Innovation.

EITAN SHAMIR: I was privileged to interview IDF senior officers who made history: MG Dan Tolkowsky, Chief of the Israeli Air Force (1953–1958); BG Yeshayahu Bareket; BG Uzi Rubin; and BG Aviam Sela. Lieutenant Colonel Moti Havakuk, Israel Air Force historian, was most helpful.

The book benefited from the help of Reichman University School of Diplomacy and Strategy students who served in the IDF in varied capacities; Eado Hecht of the IDF Staff and Command College who provided many insights; and Dima Adamsky of Reichman University. Efraim Inbar and Efraim Karsh, successive directors of the Begin-Sadat Center for Strategic Studies (BESA), provided support and helpful advice. Elad Erlich, a Bar-Ilan University graduate student, contributed his detailed knowledge of IDF research and development.

Finally, we wish to thank our editor at Harvard University Press, Kathleen McDermott, for her hard work in guiding the manuscript through the review process and offering many useful suggestions along the way.

Index